Lecture Notes in Electrical Engineering

Volume 1475

The book series *Lecture Notes in Electrical Engineering* (LNEE) publishes the latest developments in Electrical Engineering—quickly, informally and in high quality. While original research reported in proceedings and monographs has traditionally formed the core of LNEE, we also encourage authors to submit books devoted to supporting student education and professional training in the various fields and applications areas of electrical engineering. The series cover classical and emerging topics concerning:

- Communication Engineering, Information Theory and Networks
- Electronics Engineering and Microelectronics
- Signal, Image and Speech Processing
- Wireless and Mobile Communication
- Circuits and Systems
- Energy Systems, Power Electronics and Electrical Machines
- Electro-optical Engineering
- Instrumentation Engineering
- Avionics Engineering
- Control Systems
- Internet-of-Things and Cybersecurity
- Biomedical Devices, MEMS and NEMS

For general information about this book series, comments or suggestions, please contact leontina.dicecco@springer.com.

To submit a proposal or request further information, please contact the Publishing Editor in your country:

China

Jasmine Dou, Editor (jasmine.dou@springer.com)

India, Japan, Rest of Asia

Swati Meherishi, Editorial Director (Swati.Meherishi@springer.com)

Southeast Asia, Australia, New Zealand

Ramesh Nath Premnath, Editor (ramesh.premnath@springernature.com)

USA, Canada

Michael Luby, Senior Editor (michael.luby@springer.com)

All other Countries

Leontina Di Cecco, Senior Editor (leontina.dicecco@springer.com)

**** This series is indexed by EI Compendex and Scopus databases. ****

Ji Su Park · Laurence T. Yang · Yi Pan ·
James J. Park
Editors

Advanced Multimedia and Ubiquitous Engineering

Proceedings of MUE-FutureTech 2024

Springer

Editors
Ji Su Park
Department of Computer Science
and Engineering
Jeonju University
Wansan, Jeonju, Korea (Republic of)

Yi Pan
Department of Computer Science
Georgia State University
Atlanta, GA, USA

Laurence T. Yang
Department of Computer Science
St. Francis Xavier University
Antigonish, NS, Canada

James J. Park
Department of Computer Science
and Engineering
Seoul National University of Science
and Technology
Seoul, Korea (Republic of)

ISSN 1876-1100 ISSN 1876-1119 (electronic)
Lecture Notes in Electrical Engineering
ISBN 978-981-95-1564-6 ISBN 978-981-95-1565-3 (eBook)
https://doi.org/10.1007/978-981-95-1565-3

This Springer imprint is published by the registered company Springer Nature Singapore Pte Ltd.
The registered company address is: 152 Beach Road, #21-01/04 Gateway East, Singapore 189721, Singapore

MUE 2024 Organization

Honorary Chair

Young-Sik Jeong, Dongguk University, Korea

Steering Committee

James J. Park, SeoulTech, Korea

General Chairs

Jungho Kang, Baewha Woman University, Korea
Vincenzo Loia, University of Salerno, Italy

Program Chairs

Ji Su Park, Jeonju University, Korea
Fei Hao, Xi'an Normal University, China
Alireza Souri, Islamic Azad University, Iran
Sherali Zeadally, University of Kentucky, USA

International Advisory Committee

Byung Seok Shin, Inha University, Korea
Jianhua Ma, Hosei University, Japan
Shu-Ching Chen, Florida International University, USA
Han-Chieh,Chao National Ilan University, Taiwan
Weijia Jia, Shanghai Jiaotong University, China
Borko Furht, Florida Atlantic University, USA
Thomas Plagemann, University of Oslo, Norway
Roger Zimmermann, National University of Singapore, Singapore
Stephan Olariu, Old Dominion University, USA
Koji Nakano, University of Hiroshima, Japan
Houcine Hassan, Universitat Politecnica de Valencia, Spain

Publicity Chairs

Wei Song, North China University of Technology, China
Joon-Min Gil, Catholic University of Daegu, Korea
Junbo Wang, University of Aizu, Japan
Seokhong Min, Mindata Co., Korea
Byoungwook Kim, Dongguk University, Korea
Hang-Bae Chang, Chung-Ang University, Korea
Yunhee Kang, Baekseok University, Korea
Chao-Tung Yang, Tunghai University, Taiwan
Sushil Kumar Singh, SeoulTech, Korea

Program Committee

Ch. Z. Patrikakis, University of West Attica, Greece
Guillermo Camara Chavez, Universidade Federal de Minas Gerais, Brazil
Afrand Agah, West Chester University of Pennsylvania, USA
Francisco Jose Monaco, University of Sao Paulo, Brazil
Jin Kwak, Ajou University, Korea
Rainer Unland, University of Duisburg-Essen, Germany
Rajkumar Kannan, Affiliation Bishop Heber College, India
Fay Huang, National Ilan University, Taiwan
SungSuk Kim, Seokyeong University, Korea
Dongkyun Kim, KISTI, Korea
Yunhee Kang, Baekseok University, Korea
Chao-Tung Yang, Tunghai University, Taiwan

Ki-Sik Kong, Namseoul University, Korea
Xu Zhang, Chongqing University of Posts and Telecommunications, China
Waleed Farag, Indiana University of Pennsylvania, USA
Mingjie Liu, Chongqing University of Post and Telecom, China

FutureTech 2024 Organization

Honorary Chairs

Doo-soon Park, SoonChunHyang University, Korea
Qinghua Zhang, Chongqing University of Posts and Telecommunications, China

Steering Committee

James J. Park, SeoulTech, Korea
Young-Sik Jeong, Dongguk University, Korea
Rui Li, Chongqing University of Posts and Telecommunications, China

General Chairs

Jungho Kang, Baewha Women's University, Korea
Yang Xiao, The University of Alabama, USA
Changhao Piao, Chongqing University of Posts and Telecommunications, China
Sheng Lu, Chongqing University of Posts and Telecommunications, China

Program Chairs

Ji Su Park, Jeonju University, Korea
Yan Li, Inha University, Korea
Ka Lok Man, Xi'an Jiaotong-Liverpool University, China
S. Vimal, National Engineering College, India

Local Chairs

Mingjie Liu, Chongqing University of Posts and Telecommunications, China
Ping Liu, Chongqing University of Posts and Telecommunications, China
Junsheng Chen, Chongqing University of Posts and Telecommunications, China
Jianguo Miao, Chongqing University of Posts and Telecommunications, China
Junren Shi, Chongqing University of Posts and Telecommunications, China
Jiufei Luo, Chongqing University of Posts and Telecommunications, China
Chongying Deng, Chongqing University of Posts and Telecommunications, China
Xu Zhang, Chongqing University of Posts and Telecommunications, China
Yang Fang, Chongqing University of Posts and Telecommunications, China

International Advisory Committee

Victor Leung, University of British Columbia, Canada
Hsiao-Hwa Chen, National Cheng Kung University, Taiwan
Laurence T. Yang, St. Francis Xavier University, Canada
C.S. Raghavendra, University of Southern California, USA
Philip S. Yu, University of Illinois at Chicago, USA
Hai Jin, Huazhong University of Science and Technology, China
Qun Jin, Waseda University, Japan
Arun Kumar Sangaiah, VIT University, India

Publicity Chairs

Hyuk Jun Kwon, Soonchunhyang University, Korea
Jin Wang, Yangzhou University, China
Kwang-il Hwang, Incheon National University, Korea
Min Choi, Chungbuk National University, Korea
Yeong-Seok Seo, Yeungnam University, Korea
Jinho Park, Dongguk University, Korea

Sushil Kumar Singh, SeoulTech, Korea

Program Committee

Maumita Bhattacharya, Charles Sturt University, Australia
Huang Kuo-Chan, National Taichung University of Education, Taiwan
Wladyslaw Homenda, Warsaw University of Technology, Poland
Irene Chang, Shih-Hsin University, Taiwan
Yu-Chen Hu, Providence University, Taiwan
Bing Chen, Memorial University of Newfoundland, Canada
Ren-Song Ko, National Chung Cheng University, Taiwan
Joel Rodrigues, National Institute of Telecommunications, Brazil
Suren Byna, Lawrence Berkeley National Lab, USA
Abdel-Badeeh Salem, Ain Shams University, Egypt
Kyungbaek Kim, Chonnam National University, Republic of Korea
Jinli Cao, La Trobe University, Australia
Tatjana Davidovic, Serbian Academy of Sciences and Arts, Serbia
Caldelli Roberto, University of Florence, Italy

Message from the MUE 2024 General Chair

MUE 2024 is the 18th event of the series of international scientific conference. This conference takes place on April 24–26, 2024 in Chongqing, China. The aim of the MUE 2024 is to provide an international forum for scientific research in the technologies and application of Multimedia and Ubiquitous Engineering. Ever since its inception, International Conference on Multimedia and Ubiquitous Engineering has been successfully held as MUE 2023 (Phnom Penh, Cambodia), MUE 2022 (Jeju, Korea), MUE 2021 (Jeju, Korea), MUE 2020 (Jeju, Korea), MUE 2019 (Xian, China), MUE 2018 (Salerno, Italy), MUE 2017 (Seoul, Korea), MUE 2016 (Beijing, China), MUE 2015 (Hanoi, Vietnam), MUE 2014 (Zhangjiajie, China), MUE 2013 (Seoul, Korea), MUE 2012 (Madrid, Spain), MUE 2011 (Loutraki, Greece), MUE 2010 (Cebu, Philippines), MUE 2009 (Qingdao, China), MUE 2008 (Busan, Korea) and MUE 2007 (Seoul, Korea).

The conference papers included in the proceedings cover the following topics: Multimedia Modeling and Processing, Multimedia and Digital Convergence, Ubiquitous and Pervasive Computing, Ubiquitous Networks and Mobile Communications, Ubiquitous Networks and Mobile Communications, Intelligent Computing, Multimedia and Ubiquitous Computing Security, Multimedia and Ubiquitous Services, Multimedia Entertainment. Accepted and presented papers highlight new trends and challenges of Multimedia and Ubiquitous Engineering. We hope readers will find these results useful and inspiring for their future research.

We would like to express our sincere thanks to Steering Chair: James J. Park (SeoulTech, Korea). Our special thanks go to the Program Chairs: Ji Su Park (Jeonju University, Korea), Fei Hao (Xi'an Normal University, China), Alireza Souri (Islamic Azad University, Iran), Sherali Zeadally (University of Kentucky, USA), all

Program Committee members and all reviewers for their valuable efforts in the review process that helped us to guarantee the highest quality of the selected papers for the conference.

MUE 2024 General Chairs

Jungho Kang, Baewha Women's University, Korea
Vincenzo Loia, University of Salerno, Italy

Message from the MUE 2024 Program Chairs

Welcome to the 18th International Conference on Multimedia and Ubiquitous Engineering (MUE 2024) on April 24–26, 2024. MUE 2024 will be the most comprehensive conference focused on the various aspects of multimedia and ubiquitous engineering. It will provide an opportunity for academic and industry professionals to discuss recent progress in the area of multimedia and ubiquitous environment. In addition, the conference will publish high quality papers which are closely related to the various theories and practical applications in multimedia and ubiquitous engineering. Furthermore, we expect that the conference and its publications will be a trigger for further related research and technology improvements in these important subjects.

For MUE 2024, we received many paper submissions, after a rigorous peer review process, we accepted only articles with high quality for the MUE 2024 proceedings. All submitted papers have undergone blind reviews by at least two reviewers from the technical program committee, which consists of leading researchers around the globe. Without their hard work, achieving such a high-quality proceeding would not have been possible. We take this opportunity to thank them for their great support and cooperation. Finally, we would like to thank all of you for your participation in our conference, and also thank all the authors, reviewers, and organizing committee members. Thank you and enjoy the conference!

MUE 2024 Program Chairs

Ji Su Park, Jeonju University, Korea
Fei Hao, Xi'an Normal University, China
Alireza Souri, Islamic Azad University, Iran
Sherali Zeadally, University of Kentucky, USA

Message from the FutureTech 2024 General Chairs

FutureTech 2024 is the 19th event of the series of international scientific conference. This conference takes place on April 24–26, 2024 in Chongqing, China. The aim of the FutureTech 2024 is to provide an international forum for scientific research in the technologies and application of information technology. FutureTech 2024 is the next edition of FutureTech 2023 (Phnom Penh, Cambodia), FutureTech 2022 (Jeju, Korea), FutureTech 2021 (Jeju, Korea), FutureTech 2020 (Jeju, Korea), FutureTech 2019 (Xi'an, China), FutureTech 2018 (Salerno, Italy), FutureTech 2017 (Seoul, Korea), FutureTech 2016 (Beijing, China), FutureTech 2015 (Hanoi, Vietnam), FutureTech 2014 (Zhangjiajie, China), FutureTech 2013 (Gwangju, Korea), FutureTech 2012 (Vancouver, Canada), FutureTech 2011 (Loutraki, Greece) and FutureTech 2010 (Busan, Korea) which was the next event in a series of highly successful the International Symposium on Ubiquitous Applications and Security Services (UASS-09, USA, January 2009), previously held as UASS-08 (Okinawa, Japan, March 2008), UASS-07 (Kuala Lumpur, Malaysia, August, 2007) and UASS-06 (Glasgow, Scotland, UK, May, 2006).

The conference papers included in the proceedings cover the following topics: Hybrid Information Technology, High Performance Computing, Cloud and Cluster Computing, Ubiquitous Networks and Wireless Communications, Digital Convergence, Multimedia Convergence, Intelligent and Pervasive Applications, Security and Trust Computing, IT Management and Service, Bioinformatics and Bio-Inspired Computing, Database and Data Mining, Knowledge System and Intelligent Agent, Game and Graphics, and Human-centric Computing and Social Networks. Accepted and presented papers highlight new trends and challenges of future information technologies. We hope readers will find these results useful and inspiring for their future research.

We would like to express our sincere thanks to Steering Chair: James J. Park (SeoulTech, Korea), Young-Sik Jeong (Dongguk University, Korea) and Rui Li (Chongqing University of Posts and Telecommunications). Our special thanks go to the Program Chairs: Ji Su Park (Jeonju University, Korea), Yan Li (Inha University, Korea), Ka Lok Man (Xi'an Jiaotong-Liverpool University, China), S. Vimal (National Engineering College, India), all Program Committee members, and all reviewers for their valuable efforts in the review process that helped us to guarantee the highest quality of the selected papers for the conference.

We cordially thank all the authors for their valuable contributions and the other participants of this conference. The conference would not have been possible without their support. Thanks are also due to the many experts who contributed to making the event a success.

FutureTech 2024 General Chairs

Jungho Kang, Baewha Women's University, Korea
Yang Xiao, The University of Alabama, USA
Changhao Piao, Chongqing University of Posts and Telecommunications, China
Sheng Lu, Chongqing University of Posts and Telecommunications, China

Message from the FutureTech 2024 Program Chairs

Welcome to the 19th International Conference on Future Information Technology (FutureTech 2024), which will be held in Chongqing, China on April 24–26, 2024. FutureTech 2024 will be the most comprehensive conference focused on the various aspects of information technologies. It will provide an opportunity for academic and industry professionals to discuss recent progress in the area of future information technologies. In addition, the conference will publish high quality papers which are closely related to the various theories and practical applications in multimedia and ubiquitous engineering. Furthermore, we expect that the conference and its publications will be a trigger for further related research and technology improvements in these important subjects.

For FutureTech 2024, we received many paper submissions, after a rigorous peer review process, we accepted only articles with high quality for the FutureTech 2024 proceedings, published by the Springer. All submitted papers have undergone blind reviews by at least two reviewers from the technical program committee, which consists of leading researchers around the globe. Without their hard work, achieving such a high-quality proceeding would not have been possible. We take this opportunity to thank them for their great support and cooperation. Finally, we would like to thank all of you for your participation in our conference, and also thank all the authors, reviewers, and organizing committee members. Thank you and enjoy the conference!

FutureTech 2024 Program Chairs

Ji Su Park, Jeonju University, Korea
Yan Li, Inha University, Korea
Ka Lok Man, Xi'an Jiaotong-Liverpool University, China
S. Vimal, National Engineering College, India

Contents

A Novel Adaptive Coding Technology for Data Center Network

Yang Rui

Abstract Due to the increasing demand for applications by users worldwide, many service providers (such as Google and Microsoft) have deployed their services on a Cross-Data Center Network (DCN) connected by a Wide Area Network (WAN). While WAN ensures interconnection between headquarters and nodes at all levels, security issues within the network are becoming increasingly prominent. In the event of a malicious attack, data transmission across DCNs may result in packet loss, leading to various adverse outcomes such as data transmission delays, low throughput, and wastage of network bandwidth. To ensure the reliability of cross-DCN traffic transmission in the face of malicious traffic attacks, this paper proposes a coding method designed to achieve low delay and high throughput of information across DCNs during such attacks. This method utilizes Forward Error Correction (FEC) coding. The original data packets are encoded by senders into more coded packets than the original amount, allowing the receiver to recover complete valid data even in the event of packet loss. Additionally, the coding module dynamically adjusts the quantity of redundant packets to minimize the occurrence of link congestion. Through NS-3 simulations, it is demonstrated that FEC coding can reduce transmission delays by approximately 35% and improve throughput by about 30% when under attack by malicious traffic, compared to other methods, as shown by the statistics.

Keywords Data center network · FEC coding · Adaptive · Redundancy control

1 Introduction

Many service providers have invested significant capital in building data centers worldwide. Furthermore, with the increasing demand for applications from users globally, numerous service providers, such as Google and Microsoft, deploy services

Y. Rui (✉)
School of Computer & Communication Engineering, Changsha University of Science & Technology, Changsha, China
e-mail: hjshyangrui@stu.csust.edu.cn

J. S. Park et al. (eds.), *Advanced Multimedia and Ubiquitous Engineering*, Lecture Notes in Electrical Engineering 1475, https://doi.org/10.1007/978-981-95-1565-3_1

on a Cross-Data Center network connected by the WAN [1]. However, as the business volume and complexity of WAN application information systems continue to improve, security issues in the network are becoming increasingly prominent. In the event of malicious attacks, packet loss emerges as a primary concern for Cross DCN. Once packet loss occurs, the retransmission mechanism is triggered, leading to a series of adverse consequences. Therefore, achieving lower latency and higher throughput of information across DCN is crucial, especially in the face of malicious attacks.

Recent research has focused extensively on malicious attacks. One aspect of this research involves identifying, classifying, and filtering malicious attack flows, including encrypted malicious traffic [5], using various methods such as Convolutional Neural Network (CNN) [2] and Deep Neural Network (DNN) [3, 4], to mitigate attacks at their source. Another area of research involves manipulating transmitted data, employing techniques like backup [6], retransmission, and TCP tail loss recovery mechanisms [7], to duplicate data packets or transmit them multiple times to prevent data loss. However, both the methods for detecting malicious traffic and those for operating transmitted data have certain shortcomings and limitations.

In this article, we propose a method based on Forward Error Correction (FEC) coding to encode and transmit streams across data centers. This method aims to reduce the impact of packet loss caused by malicious attacks on networks across DCs and diminish the latency introduced by packet loss, thus achieving rapid convergence. The approach involves the sender using Error Correcting Code (ECC) to redundantly encode the message. Even in the event of a malicious attack, the receiver can recover the lost message as long as it receives sufficient encoded messages, eliminating the need for retransmission. This effectively avoids congestion caused by malicious attacks and delays due to data return. Additionally, the transparent coding layer only requires deployment between the TCP and IP layers of the terminal host without modifying the existing TCP/IP protocol.

The sections of this paper are organized as follows: Sect. 2 outlines the motivation for this study, Sect. 3 discusses the coding design, Sect. 4 presents the experimental results, and Sect. 5 provides the conclusion.

2 Motivation

As we all know, the Cross DCN is connected by WAN, ensuring interconnection between headquarters and nodes at all levels, thereby facilitating the full sharing and utilization of network information resources. However, due to this interconnectedness, malicious attacks in the network are increasingly prevalent.

To assess the harmfulness of malicious traffic attacks, NS-3 simulation was employed for experimentation. NS-3 simulation, compiled in C++, is a network simulator used to model network topology and execute various network protocols, thereby simulating the operational mechanisms of computer networks. Experiments were conducted on a double Leaf-Spine topology network, as depicted in Fig. 1. This network comprises 2 hosts, 2 leaf switches, 3 spine switches, and 1 central switch. The double Leaf-Spine topology represents two data centers, while the central switch and its connected links symbolize the WAN connecting these data centers. In the double Leaf-Spine topology, each switch has a buffer size of 256 packets. Multiple equal-cost paths exist between source and target nodes, with each path having a bandwidth of 1 Gbps and a round-trip propagation delay of 100 μs.

To simulate packet loss resulting from malicious traffic attacks, senders S1, S2, ..., S4 transmit data packets to receivers C1, C2, ..., C4, while sender S0 concurrently sends packets to receivers randomly for malicious attacks. The attack intensity, representing the proportion of malicious traffic to effective traffic, gradually increases from 0 to 40 percent. As illustrated in Fig. 2, the average throughput of receivers decreases as the intensity of malicious attacks increases.

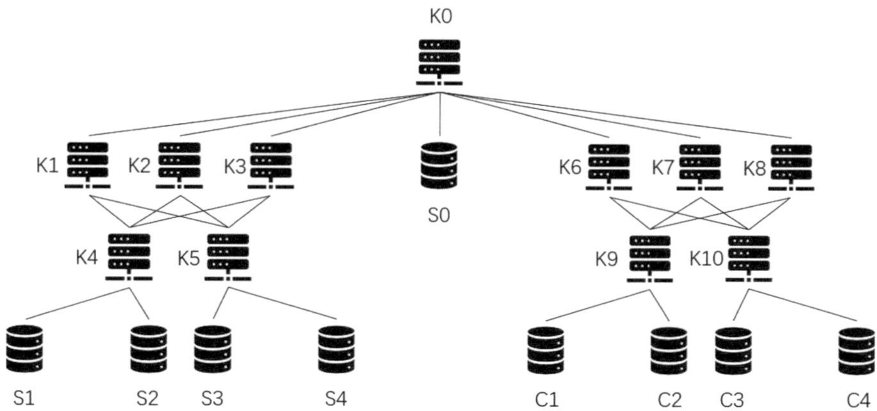

Fig. 1 Double leaf-spine topology

Fig. 2 Relationship between throughput and attack intensity

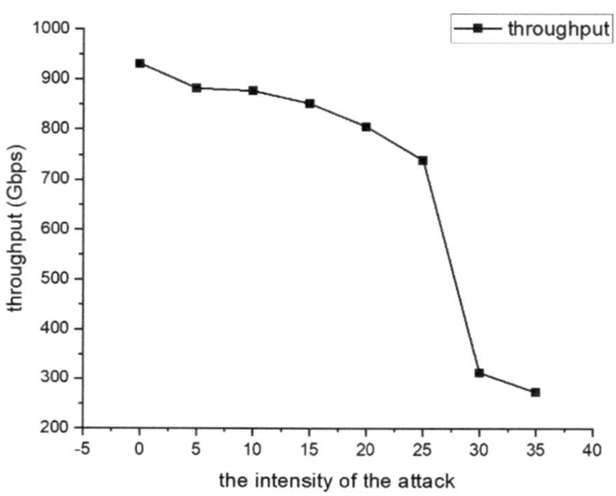

3 System Design

3.1 Design Overview

FEC can timely recover lost packets by employing redundant packets, thereby enhancing transmission performance. Generally, adding a reasonable number of redundant packets can improve transmission reliability. Specifically, in the event of a malicious attack in the Cross DCN leading to the loss of effective packets, the FEC coding method with redundant packets does not immediately trigger the retransmission mechanism. As long as the number of lost packets is smaller than the number of redundant ones, the original packets can be recovered, enabling the receiver to successfully receive the data. The framework of FEC coding is depicted in Fig. 3.

For senders, the source packets are passed to the encoding module for encoding. Within this module, an ACK is sent to notify the sender of successful receipt of the source packet. Upon receiving the ACK packet, the sender removes it and forwards it to the transmission layer. Subsequently, the senders generate $k + r$ encoded data packets through operations such as XOR summation on the k original data packets within the encoding module before transmitting them. It's noteworthy that in the coding process, r is not fixed, indicating that the number of redundant packets dynamically adjusts based on the real-time state of network traffic.

For receivers, the decoding module receives coded packets from senders and decodes $k + s$ (where $s <= r$) coded packets. After successful decoding, the decoded

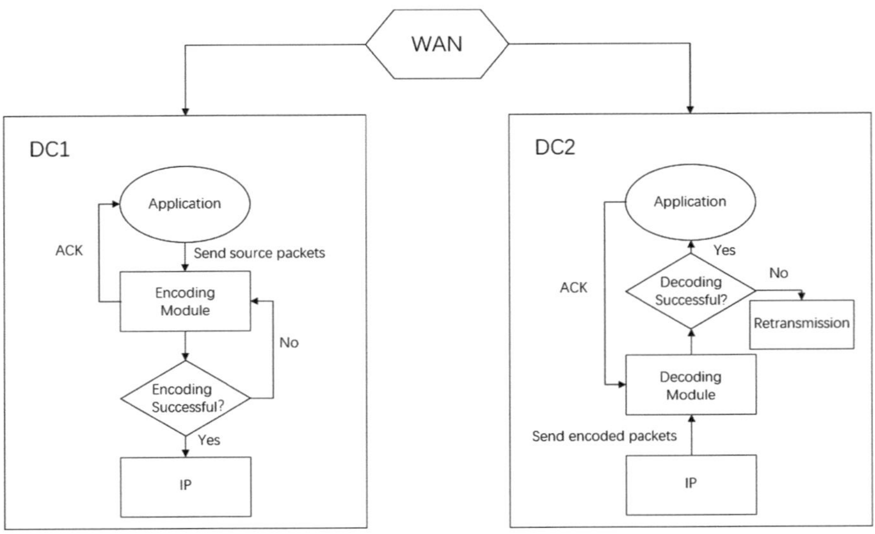

Fig. 3 FEC coding framework

packets are handed over to the upper layer. Simultaneously, receivers send an ACK to confirm the successful reception of the packet.

3.2 Design Details

Before sending data, the sender needs to partition the transmitted stream into m groups of equal length. Each group is then further divided into k packets of equal size. Subsequently, the sender evaluates the packet receipt rate pi (where i=1, ..., m) in the WAN by sending test data packets. Each group of packets sent must undergo testing for the packet receipt rate, requiring a total of m evaluations. For each packet, r1, r2, ..., rm redundant packets are generated respectively. Since the necessary condition for successful decoding is that the number of data packets received by the receiver is greater than or equal to the number of source data packets, ri must satisfy the condition $(k + ri)\ pi >= k$ to ensure that, on average, the receiver can receive encoded data packets equal to or greater than the number of source data packets.

The coding module of the sender generates redundant packets through the source packets, created by performing the XOR operation on the source packets. During transmission, in addition to sending the source data packets individually, the sender also transmits these redundant data packets. The Equal-Cost Multi-Path (ECMP) mechanism is deployed on leaf switches to ensure load balancing across each link and enhance link utilization. The decoding module of receivers utilizes the Belief Propagation (BP) algorithm to decode the received encoded data packets.

4 Experimental Results and Analysis

4.1 Experimental Setup

NS-3 simulation was chosen for this experiment, conducted in the topology depicted in Fig. 4. Switch K0 interconnects two Leaf-Spine topology networks, each representing a data center, while switch K0 and its connected links represent the WAN connecting these data centers. In this setup, S1–S6 serve as senders from the data center, while C1–C6 act as receivers. Specifically, S1 sends data to C1, S2–C2, and so forth. Each leaf switch employs the ECMP mechanism to evenly distribute received packets across each link. Additionally, S0 randomly sends packets to C1–C6 for malicious attacks. All other settings remain consistent with the motivation.

The attack intensity varies periodically and randomly from 0 to 40 to simulate packet loss resulting from malicious traffic attacks. The period of this intensity variation is set to 1000 μs. Each sender transmits a 1000 MB stream to its respective receiver. Before transmission, the stream from each sender is divided into 100 groups, each containing 10 packets.

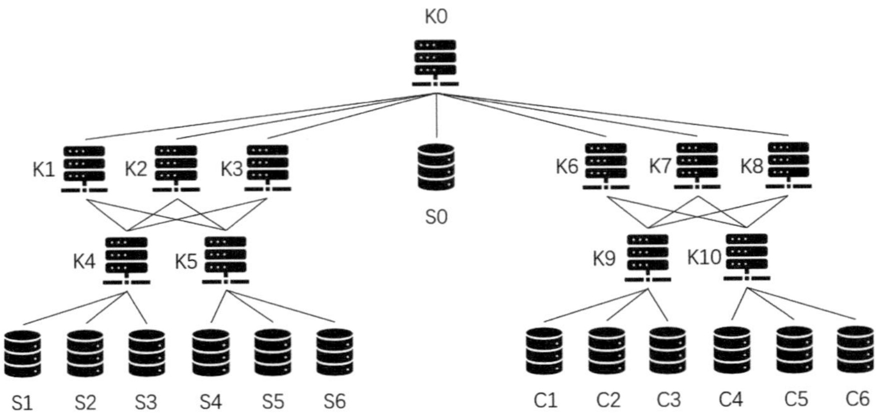

Fig. 4 Cross-data center topology

During data collection, two metrics are selected: Flow Completion Time (FCT) and Receiver Throughput. To calculate these metrics, the FCT of the 6 flows and the throughput of the 6 receiver hosts are obtained. The average values of these two metrics are then calculated to derive an overall indicator.

4.2 Experimental Results

Figure 5 displays the FCT of FEC, DCTCP, GTCP, and Gemini. In the figure, the FEC method proposed in this article exhibits the shortest FCT, averaging about 35% less than the other three methods. This suggests that FEC effectively reduces traffic transmission delays in scenarios involving malicious attacks across data center networks, surpassing the performance of the existing three methods.

Figure 6 illustrates the throughput of FEC coding in comparison with DCTCP, GTCP, and Gemini methods. It is evident that the throughput of FEC coding is the highest among the four methods, exhibiting an increase of approximately 20% to 30%. The fluctuations observed may be attributed to the attack intensity randomly set by host S0.

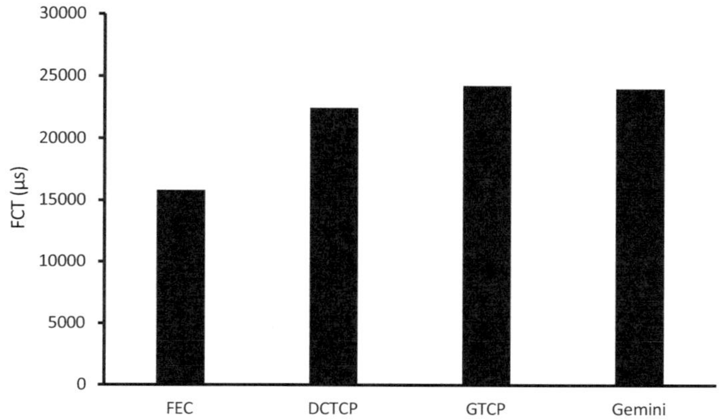

Fig. 5 FCT of various methods

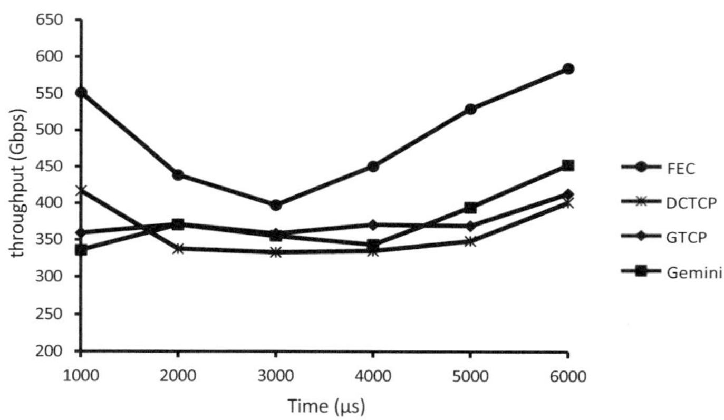

Fig. 6 Throughput of various methods

5 Conclusion

This paper introduces the research on FEC coding in the context of malicious attacks on cross-data center traffic transmission. FEC coding is an adaptive technology that dynamically adjusts redundancy settings in response to network conditions, enabling the receiver to successfully recover source data even without receiving complete data. Due to its ability to make real-time adjustments based on network conditions, FEC coding meets decoding requirements while minimizing unnecessary bandwidth waste and avoiding additional congestion caused by redundant data packets. Moreover, FEC coding maintains compatibility with existing transport layer protocols and does not require complex hardware modifications during deployment. Experimental results demonstrate the advantages of adaptive FEC coding technology over existing methods. In future work, FEC coding can explore integrating real-world network environments and business requirements for more comprehensive simulation testing and field deployment verification. Additionally, combining knowledge from other fields to obtain network status information can enable FEC to adjust redundancy settings more efficiently in real time. Furthermore, addressing the grouping problem in FEC coding preprocessing, Artificial Intelligence (AI) and Machine Learning (ML) technologies can be considered to optimize FEC coding, identifying the most suitable packet grouping for the current network and enabling more intelligent and adaptive network transmission control.

References

1. Wang, Z., Li, Z., Liu, G., et al.: Examination of WAN traffic characteristics in a large-scale data center network. In: Proceedings of the 21st ACM Internet Measurement Conference, pp. 1–14 (2021)
2. Choudhury, N., Deka, D., Tewari, A., et al.: Malicious traffic classification using convolutional neural network. In: 2023 14th International Conference on Computing Communication and Networking Technologies (ICCCNT). IEEE, pp. 1–7 (2023)
3. Zhou, Y., Shi, H., Zhao, Y., et al.: Identification of encrypted and malicious network traffic based on one-dimensional convolutional neural network. J. Cloud Comput. **12**(1), 53 (2023)
4. Anitha, T., Aanjankumar, S., Poonkuntran, S., et al.: A novel methodology for malicious traffic detection in smart devices using BI-LSTM–CNN-dependent deep learning methodology. Neural Comput. Appl. **35**(27), 20319–20338 (2023)
5. Wang, Z., Fok, K.W., Thing, V.L.L.: Machine learning for encrypted malicious traffic detection: approaches, datasets and comparative study. Comput. Secur. **113**, 102542 (2022)
6. Das, S., Panda, K.G., Sen, D., et al.: Maximizing last-minute backup in endangered time-varying inter-datacenter networks. IEEE/ACM Trans. Netw. **29**(6), 2646–2663 (2021)
7. Rajiullah, M., Hurtig, P., Brunstrom, A., et al.: An evaluation of tail loss recovery mechanisms for TCP. ACM SIGCOMM Comput. Commun. Rev. **45**(1), 5–11 (2015)

Research on Anti-Interrupted Sampling Repeater Jamming Method Based on Joint Matched Filter Design for Random Subcarrier Frequency Coding Time–Frequency Encoded Multi-Carrier Signals

Ji Li and Rui Yan

Abstract Interrupted Sampling Repeater Jamming (ISRJ) is a sophisticated radar interference technique. This paper presents a countermeasure involving Random Subcarrier Frequency Coding Time–Frequency Encoded Multicarrier Signals (RSFC-MCPC) and matched filters, enhancing signal stealth and interference resistance. The approach combines waveform and filter design, employing chaotic coding for signal security at the transmitter and matched filtering for echo processing at the receiver. This method effectively suppresses interference, especially in complex scenarios with overlapping signals, achieving an average 5.5 dB improvement in signal-to-jammer ratio (SJR) over conventional MCPC techniques.

Keywords Random subcarrier frequency coding time–frequency encoded · Matched filter design · Radar signal processing · Interrupted sampling repeater jamming suppression

1 Introduction

In the advanced electronic warfare field, rapid technological advancements challenge reconnaissance and countermeasures [1]. ISRJ is a new type of coherent jamming derived from DRFM [2]. It works by selectively sampling enemy radar signals and retransmitting them to confuse and deceive enemy radars [3].

Research in ISRJ suppression encompasses both waveform design and filter strat-egies. Su Han Ning's team [4] focused on signal complementarity through phase-encoded waveforms for ISRJ distinction. Techniques such as Time–Frequency

J. Li · R. Yan (✉)
School of Computer and Communication Engineering, Changsha University of Science & Technology, Changsha, Hunan, China
e-mail: yanrui11q@126.com

J. S. Park et al. (eds.), *Advanced Multimedia and Ubiquitous Engineering*, Lecture Notes in Electrical Engineering 1475, https://doi.org/10.1007/978-981-95-1565-3_2

Random Coding (TFRC) [7] and the integration of MIMO radar technology [8] have shown significant improvements in jamming resistance. In filtering, Gao Yu Hang et al. [5] used energy constraints to isolate key signals for interference removal. Another study [6] used mismatched filter design to effectively suppress ISRJ, improving anti-jamming performance through design and algorithm enhancements. Collectively, these advancements enhance the detection and suppression of ISRJ.

2 Analysis of ISRJ Characteristics

ISRJ includes Interrupted Sampling Direct Repeater Jamming (ISDRJ), Periodic Re-peater Jamming (ISPRJ), and Cyclic Repeater Jamming (ISCRJ), differing in re-transmission approaches. ISDRJ samples and immediately retransmits radar signal fragments within a pulse cycle. ISPRJ samples a signal segment, retransmits it multi-ple times before sampling anew, and repeats until the pulse ends. ISCRJ samples radar signals, splices, and retransmits different fragments after a delay, impacting jamming effectiveness variably. A diagram illustrating ISRJ principles is noted as Fig. 1.

The interrupted sampling rectangular envelope pulse train can be expressed as:

$$p(t) = rect\left(\frac{t}{\tau}\right) \otimes \sum_{n=-\infty}^{\infty} \sigma(t - nT_S) \tag{1}$$

τ represents the sampling pulse width, T_S denotes the sampling period, $D_r = \frac{t}{T_S}$ is defined as the duty cycle, $\otimes$ symbolizes convolution operation, and the functional expression of $rect(x)$ is given by:

$$rect(x) = \begin{cases} 1, & |x| \leq 1/2 \\ 0, & |x| > 1/2 \end{cases} \tag{2}$$

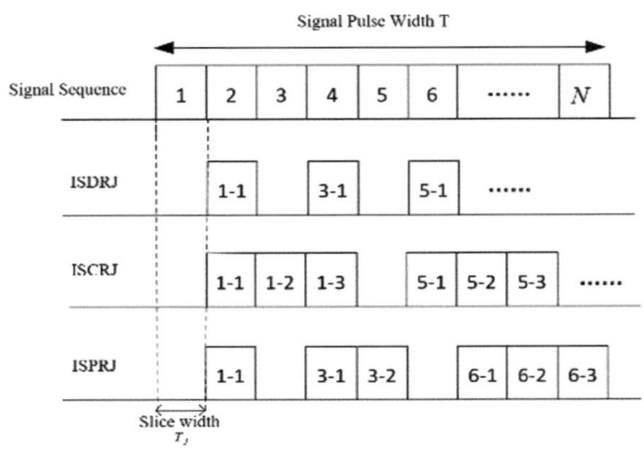

Fig. 1 Schematic diagram of interrupted sampling repeater jamming

3 Jamming Suppression Methods

3.1 RSFC-MCPC Signal Modelling

This section introduces RSFC-MCPC, enhancing anti-jamming by utilizing mutual cover among subcarriers and applying chaos and redundancy coding to increase signal complexity. This approach effectively suppresses interference, conceals the signal, and demonstrates strong anti-jamming performance against ISRJ.

The complex envelope expression of the RSFC-MCPC signal is shown in Eq. (3).

$$\begin{aligned} s(t) &= \sum_{n=1}^{P} \omega_n \partial_n \exp(j2\pi f_p t)\mu_n(t) \\ &= \sum_{n=1}^{P}\sum_{m=1}^{M} \omega_n \partial_n \exp(j2\pi f_n t)\mathrm{rect}[t-(m-1)]t_b \end{aligned} \tag{3}$$

In the formula, P represents the number of subcarriers, M denotes the number of chips for the subcarrier code $\mu_n(t)$ signifies the complex envelope of the nth subcarrier, and $\omega_n = |\omega_n|e^{j\varphi_{n,m}}$ indicates the amplitude and the weighted phase coefficient of the nth subcarrier. $\partial_n = \exp(j\varphi(n,m)t)\mathrm{rect}_n(m)$ is the chaotic phase coding value on the mth chip of the nth subcarrier. The $\mathrm{rect}_n(m)$ function defines the fixed interval of the cover symbol. $f_n = (n-1)\Delta f$ represents the subcarrier frequency of the nth subcarrier, and $\Delta f = \frac{1}{t_b}$ stands for the frequency interval between subcarriers. The $\mathrm{rect}_n(m)$ function can be expressed as:

$$rect_n(m) = \begin{cases} 1, & (n-1)*(M/P) < m \le n*(M/P) \\ 0, & \text{else} \end{cases} \tag{4}$$

The time–frequency structure of the RSFC-MCPC signal is shown in Fig. 2.

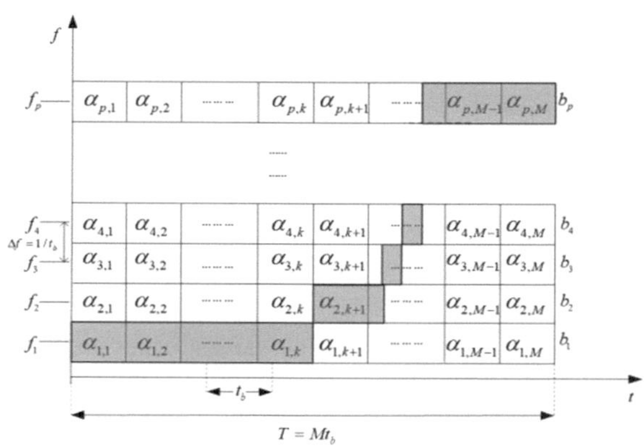

Fig. 2 RSFC-MCPC time–frequency structure

The distance between the jammer and the radar is R_j, the ISRJ signal can be expressed as follows:

$$S_j(t) = \sum_{k=0}^{K_s}\sum_{n=1}^{P}\sum_{m=1}^{M} \omega_n \partial_n \exp(j2\pi f_n(t - \Delta\tau(k)))\mathrm{rect}\left[\frac{t - \Delta\tau(k)}{\tau}\right] \tag{5}$$

In the equation, $\Delta\tau(k) = \frac{2R_j}{c} + kT_s$ calculates the radar-to-jammer delay, with k being the sampling instance count and K_s the total occurrences.

3.2 Optimized Design of Matched Filter

This paper presents a novel matched filter design for ISRJ, diverging from the conventional approach where the impulse response h(t) is the time-reversed conjugate of s(t), to using a modified signal $s'(t)$ for $h(t) = s'^*(-t)$. This method ensures subcarriers still match well despite ISRJ contamination, leading to some mismatching in the echo data. The pulse response coefficients of the filter are:

$$\begin{aligned} h(t) = s'^*(-t) &= \sum_{n=1}^{P} \omega_n \beta_n \exp\left(-j2\pi f_p t\right)\mu_n(t) \\ &= \sum_{n=1}^{P}\sum_{m=1}^{M} \omega_n \beta_n \exp\left(-j2\pi f_p t\right)\mathrm{rect}[t + (m-1)t_b] \end{aligned} \tag{6}$$

$$\mathrm{rect}'_n(m) = \begin{cases} 1, & m = n * (M/P) \\ 0, & \text{else,} \end{cases} \tag{7}$$

The formula uses $\beta_n = \exp(-j\varphi(n, m)t)\mathrm{rect}'_n(m)$ for chaotic phase coding on subcarrier chips, with $\mathrm{rect}'_n(m)$ marking the interval. The echo $S_r(t)$ and matched filter output follow:

$$y(t) = \int_{-\infty}^{\infty} S_r(\tau)h(t - \tau)d\tau = \int_{-\infty}^{\infty} S_r(\tau)s'^* * (-t + \tau)d\tau \tag{8}$$

3.3 Anti-Jamming Implementation Process

The specific process for implementing the Anti-Jamming Algorithm through the Combined Design of RSFC-MCPC Signal and Matched Filter is shown in the diagram. The detailed implementation process is as follows (Fig. 3):

The specific process is introduced as follows:

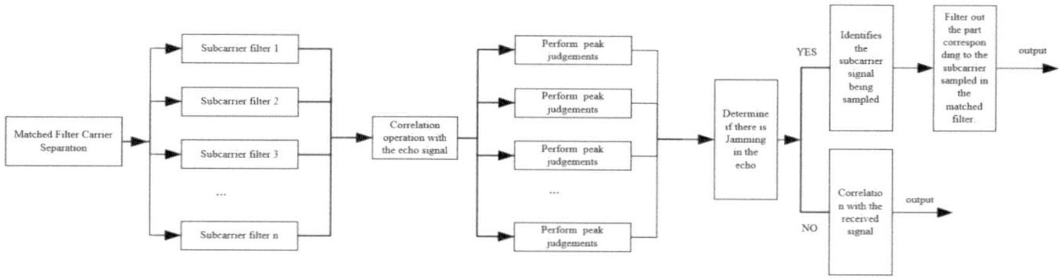

Fig. 3 Anti-jamming algorithm process

1. Generate a cover signal s/(t) from s(t), apply a Fourier transform for subcarrier separation, and create filter pulse responses for P subcarriers;
2. Perform match filtering on the echo signal's subcarriers using the filter bank, obtaining P results;
3. Peak detection analyzes Pth matched filter results to find the highest peak y_{subc} and its amplitude $\mathrm{MAX}\left[\mathrm{E}_{y_{subc}}\right]$. It then checks all P results for peaks z_{subc} within 2dB of y_{subc}. A single qualifying peak indicates no interference; multiple peaks suggest interference, marking the subcarrier;
4. If no subcarriers are marked, proceed with normal matched filtering.
5. If interference is detected in a subcarrier, its band is removed from the matched signal. Matched filtering, minus the affected subcarrier, is then applied to the echo, yielding results free from interference.

4 Simulation Results and Analysis

4.1 Anti-ISRJ Experiments and Analyses

The simulation software used is MATLAB R2018a. Simulation tests on the method's resistance to ISDRJ, ISPRJ, and ISCRJ jamming involve a radar signal with a 200 MHz bandwidth, 20us pulse width, 500 chips, 35 GHz carrier frequency, and 8 subcarriers, against a 4us jammer sampling width. Conducted under conditions of − 2 dB SNR and − 10 dB SJR, these simulations validate the anti-jamming effectiveness.

The effectiveness of the proposed suppression method against ISDRJ, ISPRJ, and ISCRJ jamming is demonstrated through simulation results in Fig. 4. Initially, all three types of jamming create realistic false targets that hinder target detection. After applying the suppression techniques, interference is significantly reduced across the board: ISDRJ's false targets are minimized, ISPRJ's multiple false targets are attenuated, and the complex false targets created by ISCRJ are also substantially lowered in amplitude. These improvements allow for the clear identification of real targets, proving the method's broad effectiveness in counteracting intermittent sampling jamming and enhancing target signal clarity.

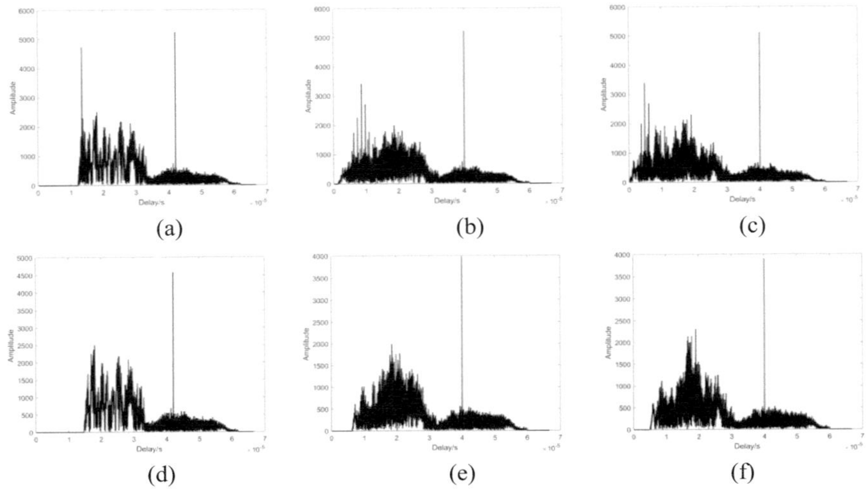

Fig. 4 Radar pulse compression under ISRJ, **a**, **b**, and **c** represent the pre-pulse compression results of ISDRJ, ISPRJ, and ISCRJ, respectively, while **d**, **e**, and **f** show the post-suppression pulse compression results

4.2 Evaluation of Jamming Suppression Performance

Our study highlights the superiority of our ISRJ suppression method against the original MCPC and techniques from references [7, 8], by conducting 500 Monte Carlo simulations with SJRs of – 6 and – 9 dB across different SNRs. Figure 5 reveals that our method excels in interference identification and suppression, showing low sensitivity to varying input SNRs.

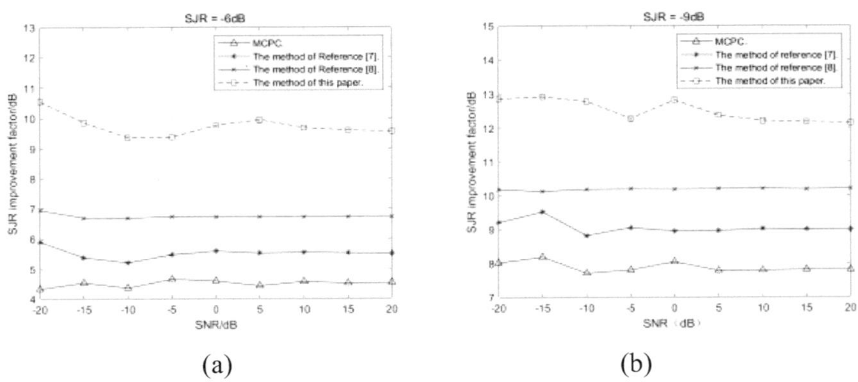

Fig. 5 Performance comparison of different methods, **a** SJR = – 6 dB, **b** SJR = – 9 dB

5 Conclusion

This paper introduces an anti-jamming strategy against ISRJ using RSFC-MCPC and matched filters, effectively countering ISRJ's threat by optimizing signal and filter design. Despite a slight impact on target accuracy, the method significantly enhances Signal to Jammer Ratio, proving its value. Future work will aim to further improve this approach, focusing on reducing its effect on target signals and increasing anti-jamming efficacy in complex environments.

Acknowledgements This work was supported under the Hui Yan Xing Dong (62402010206).

References

1. He, J., Cheng, Z., He, Z.: Joint design of detecting and screening waveforms against interrupted sampling repeater jamming with inaccurate jamming knowledge. Digit. Signal Process. **130**, 103679 (2022)
2. Han, B., Qu, X., Yang, X., Zhang, Z., Li, W.: DRFM repeater jamming suppression method based on joint range-angle sparse recovery and beamforming for distributed array radar. Remote Sens. **15**, 3449 (2023)
3. Wang, X., Li, B., Liu, W., Chen, H., Zhu, Y., Ni, M.: Anti-interrupted sampling repeater jamming based on intra-pulse frequency modulation slope agile radar waveform joint FrFT. Digit. Signal Process. **147**, 104418 (2024)
4. Su, H. et al.: Waveform-domain complementary signal sets for interrupted sampling repeater jamming suppression. IEEE Trans. Aerosp. Electron. Syst. **60**(5):7142–58 (2024)
5. Gao, Y., Fan, H., Ren, L., Liu, Z., Liu, Q., Mao, E.: Joint design of waveform and mismatched filter for interrupted sampling repeater jamming suppression. IEEE Trans. Aerosp. Electron. Syst. **59**(6), 8037–8050 (2023)
6. Zhou, K., Su, Y., Wang, D., Ma, H., Liu, L., Li, C.: Improved SAR interrupted-sampling repeater jamming countermeasure based on waveform agility and mismatched filter design. IEEE Trans. Geosci. Remote Sens. **61**, 1–16 (2023)
7. Li, J., Luo, X., Duan, X., et al.: A novel radar waveform design for anti-interrupted sampling repeater jamming via time-frequency random coded method. Prog. Electromagn. Res. M **98**, 89–99 (2020)
8. Li, J., Deng, Q., Ou, J., Wang, W.: Research on multi-carrier phase code signal against interrupted sampling repeater jamming based on MIMO radar. J. Signal Process. **38**(7), 1488–1496 (2022)

Waveform Design of OFDM-SF for Sea Clutter Suppression Based on Joint Filters

Ji Li and JiaLiang Li

Abstract Sea clutter is influenced by numerous complex factors, particularly in grazing incidence scenarios, where it significantly affects radar performance in detecting surface targets, creating unavoidable interference. This paper analyzes the anti-interference performance of different signals against sea clutter. From the perspective of waveform generation, an Orthogonal Frequency Division Multiplexing-Subband Filter (OFDM-SF) signal is designed. Experimental results show that OFDM-SF outperforms conventional OFDM radar, achieving a 3 dB improvement in signal-to-clutter ratio when combating sea clutter.

Keywords OFDM radar · Sea clutter · Clutter suppression

1 Introduction

The rapid development of communication technology has led to a growing complexity in the electronic warfare environment. During ocean detection, various forms of interference, such as passive and active interference, present challenges, with sea clutter interference being particularly severe [1]. There is currently extensive research on sea clutter and anti-sea clutter techniques. By analyzing sea clutter through simulation, researchers can grasp its characteristics and devise strategies for interference mitigation [2, 3]. Moreover, several algorithms have been proposed to counteract the impact of sea clutter or extract target information [4, 5]. These advancements not only improve radar capabilities in oceanic environments but also inspire new ideas for future research and applications.

In this paper, a method for suppressing sea clutter interference in OFDM-SF radar signals is proposed, building upon traditional OFDM radar signals[6][7]. The approach involves introducing subcarriers into the OFDM transmission signal and employing filters during the reception of echoes. The addition of subcarriers enhances

J. Li · J. Li (✉)
School of Computer and Communication Engineering, Changsha University of Science & Technology, Changsha, Hunan, China
e-mail: ljlhnny@163.com

J. S. Park et al. (eds.), *Advanced Multimedia and Ubiquitous Engineering*, Lecture Notes in Electrical Engineering 1475, https://doi.org/10.1007/978-981-95-1565-3_3

the utilization of signal bandwidth and reduces the correlation between signals at different frequencies. The inclusion of corresponding subcarrier filters at the echo reception end not only filters out clutter outside the bandwidth but also effectively reduces the correlation between clutter within the bandwidth and echo signals. This modification allows the radar signal to effectively suppress interference from sea clutter.

2 Anti-Sea Clutter Methods Based on OFDM Radar

2.1 Radar Received Signal Model

Assuming the radar transmission signal is an OFDM-SF signal, its time-domain expression is:

$$S(t) = \sum_{k=0}^{M-1} \sum_{k=0}^{N-1} B_k e^{j(2\pi f_k t + \varphi_k)} \tag{1}$$

In the expression, B_k, f_k, and φ_k represent the complex input data, frequency, and initial phase of the k-th subcarrier of the OFDM-SF radar signal, respectively; N is the number of subcarriers, and M is the number of subbands.

The target echo signal located at a distance R_0 is:

$$S_{R_0}(t) = A * S(t - t_0) = A * \sum_{k=0}^{M-1} \sum_{k=0}^{N-1} B_k e^{j(2\pi f_k t + \varphi_k)} \tag{2}$$

where A is the target echo attenuation amplitude, $t_0 = \frac{2R_0}{c}$ is the one-way propagation delay from the radar to the target, and c is the speed of electromagnetic wave propagation, i.e., the speed of light.

When the distance between the radar and the sea surface is R_1, the interference signal from the sea surface echo is:

$$S_{R_1}(t) = A * S(t - t_1) = A * \sum_{k=0}^{M-1} \sum_{k=0}^{N-1} B_k e^{j(2\pi f_k t + \varphi_k)} \otimes F_K \tag{3}$$

In the equation, $t_1 = \frac{2R_1}{c}$ is the one-way propagation delay from the radar to the sea surface, where c is the speed of light. The symbol $\otimes$ represents convolution, and F_K is the sea clutter model.

Therefore, the received signal can be described as:

$$S_{echo}(t) = S_{R_0}(t) + S_{R_1}(t) + n(t) \tag{4}$$

where n(t) is Gaussian white noise.

Typically, the target and the sea surface are located at different positions, and the propagation delays of the pulse signals they return are also different. Therefore, the received radar signal is essentially the superposition of the target echo signal and interference signal at different time intervals.

2.2 Radar Received Signal Model

Differing from common interference mitigation methods for OFDM waveforms [8], the proposed OFDM-SF radar signal approach in this paper originates from the generation of radar signals and the reception of echoes. Specifically designed for noise and sea clutter in high-sea state scenarios, it employs synchronous filtering for noise and clutter in conjunction with target echo signals. This is followed by motion compensation and signal processing techniques such as pulse compression, aiming to achieve anti-sea clutter capabilities.

The modulation process flowchart of the OFDM-SF signal is illustrated in Fig. 1. In the actual signal generation process, to expedite the processing speed of the OFDM-SF system, fast Fourier transform (FFT) and inverse fast Fourier transform (IFFT) are commonly employed for the modulation and demodulation of OFDM-SF signals.In the OFDM-SF system, the time-domain sampled signal is composed of N subbands, each subband consisting of several subcarriers, with each subband exhibiting different frequency characteristics. The generated baseband sequence undergoes serial-to-parallel conversion, aiming to achieve multi-stream separation of data flows. After converting the serial data flow into a parallel data flow, encoding mapping is performed. This paper employs Binary Phase Shift Keying (BPSK) modulation, enhancing its resistance to interference while also improving the spectral efficiency of the system. The mapped signal then undergoes Inverse Fast Fourier Transform (IFFT) and parallel-to-serial conversion, transforming the frequency-domain signal into a time-domain signal. These processed signals are then superimposed to form a complete OFDM-SF transmitted signal.

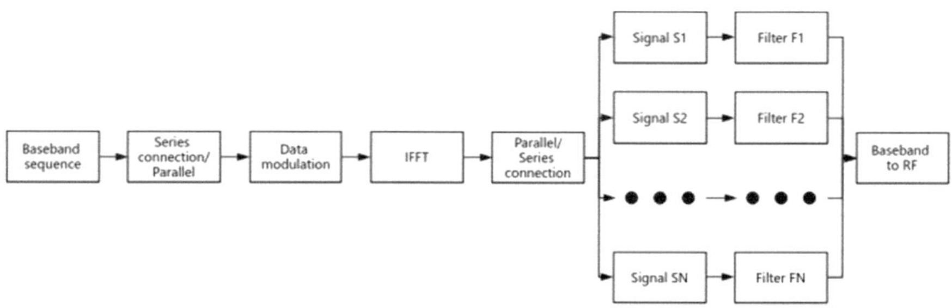

Fig. 1 OFDM-SF modulation flow block diagram

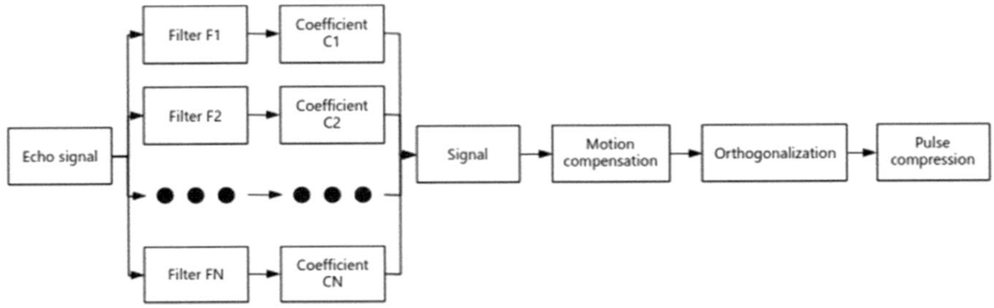

Fig. 2 Flow chart of echo signal processing

In the echo processing, based on the OFDM system model, this paper has designed an interference suppression model algorithm flowchart as shown in Fig. 2:

(1) For the input target echo signal and interference signal, as each subband has its own frequency domain, the first step is to perform filtering N times, resulting in N filtered signals;
(2) For the N filtered signals, individual coefficients are set, and the signals are synthesized.
(3) After obtaining the synthesized signal, motion compensation, intermediate frequency orthogonalization, and pulse compression processing are applied to search for the target information.

To evaluate the performance of the proposed anti-sea clutter interference method in this paper, the Signal-to-Clutter Ratio Improvement (τ_{SCR}) are introduced, expressed as:

$$\tau_{SCR} = SCR_{PC} - SCR \tag{5}$$

SCR represents the ratio of the signal to sea clutter, and SCR_{PC} is the SCR value after pulse compression.

3 Anti-Sea Clutter Methods Based on OFDM Radar

In this section, the experiment simulation was conducted using OFDM-SF radar waveform, OFDM radar waveform, and LFM radar waveform. Comparative experiments were performed to validate the superior anti-sea clutter capability of the OFDM-SF radar waveform.

Where the OFDM-SF radar transmission signal has a bandwidth of 120 MHz, a carrier frequency of 30 GHz, 8 subcarriers, 10 subbands, and a pulse width of 8 microseconds.

The OFDM radar transmission signal has a bandwidth of 120 MHz, a carrier frequency of 30 GHz, 8 subcarriers, and a pulse width of 8 microseconds. The LFM radar transmission bandwidth is 120 MHz, with a carrier frequency of 30 GHz and a pulse width of 8 microseconds. The simulation incorporates a 5-level K-distributed sea state and adds 10 dB Gaussian white noise. The radar transmission platform is located at coordinates [0,0,1600], and the target is positioned at [2000,0,10]. A schematic diagram is provided in Fig. 3:

Echo Profiles of the Three Signals:

In Fig. 4a, the time-domain echo profile of the LFM signal displays complex fluctuations, possibly attributed to multipath effects or target reflections. In the frequency-domain image, signal energy is concentrated, maintaining a certain structure.

In Fig. 4b, the time-domain echo profile of the OFDM signal exhibits higher noise, potentially interfering with target identification. However, the frequency-domain

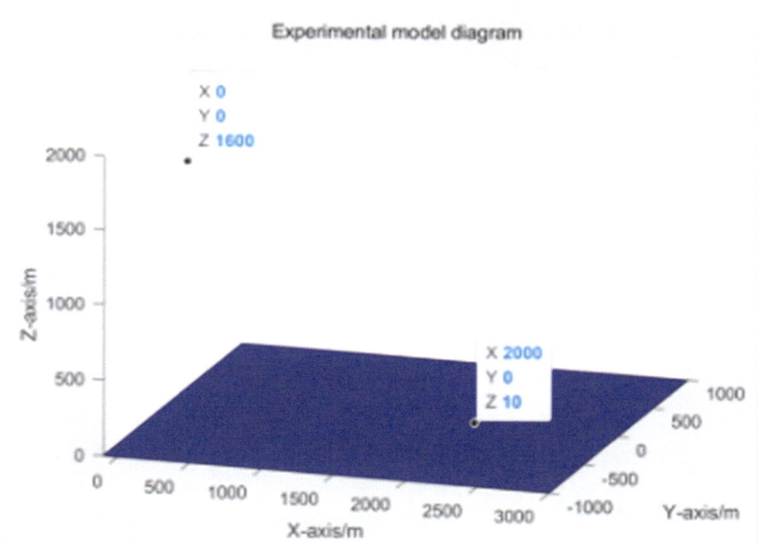

Fig. 3 Experimental model diagram

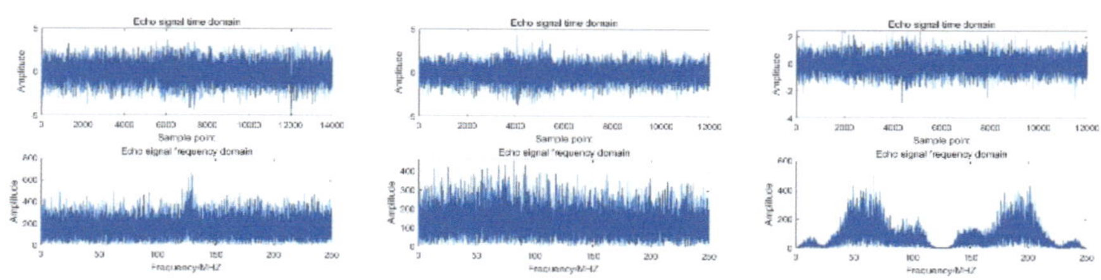

(a) (b) (c)

Fig. 4 Echo pattern

image shows multiple peaks, indicating the ability to distinguish targets through subcarriers even in the presence of noise.

In Fig. 4c, the time-domain signal after filtering for the OFDM-SF signal shows relatively uniform amplitude variations. The filtering process effectively reduces high-frequency noise, improving the signal-to-noise ratio (SNR) and signal-to-clutter ratio (SCR). This is advantageous for target identification, reducing noise and sea clutter, aiding target detection algorithms in recognizing genuine target echoes.

Pulse Compression Images of the Three Signals:

In Fig. 5a, the upper part of the image shows the LFM signal after pulse compression, displaying a distinct narrow pulse peak. The high amplitude indicates a higher signal-to-noise ratio relative to the noise background. In the lower part of the image, aside from the main echo peak, the amplitudes in other regions are significantly lower. This suggests that pulse compression effectively enhances the signal-to-noise ratio, facilitating target identification in the presence of noise and sea clutter.

In Fig. 5b, the upper part of the image shows the OFDM signal after pulse compression, revealing a prominent peak. In the lower part of the image, the main peak is clearly visible above the noise and sea clutter.

In Fig. 5c, the upper part of the image shows the OFDM-SF signal after pulse compression, revealing a highly prominent peak representing the target echo. This is advantageous for target localization. The amplitude of this peak is significantly higher than any other signals, indicating a high distinguishability between the target echo and background noise after pulse compression processing. In the lower part of the image, the signal levels around the peak are notably lower, demonstrating effective resistance to interference from noise and sea clutter (Table 1).

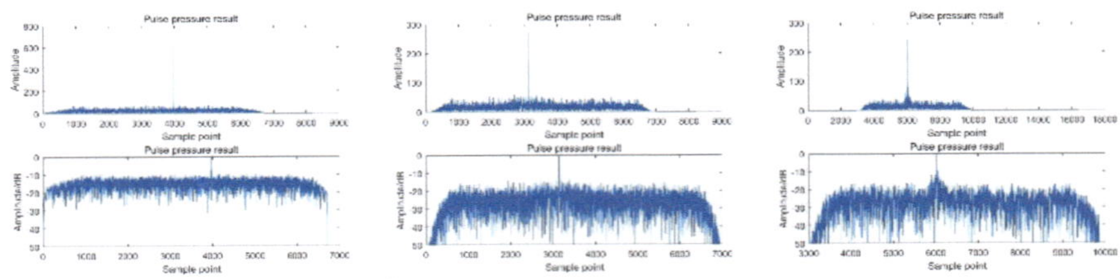

(a) (b) (c)

Fig. 5 Compressed pulse image

Table 1 Three kinds of signal SCR values

Signal	SCR (dB)	SCR_{PC} (dB)	τ_{SCR} (dB)
LFM	9.07	17.70	8.63
OFDM	9.27	18.53	9.26
OFDM-SF	9.62	22.20	12.58

4 Conclusion

The OFDM-SF radar signal based on the proposed joint filter in this paper, incorporating subcarrier design during signal generation and subcarrier filter settings at the echo reception end, effectively suppresses sea clutter interference. It achieves a signal-to-clutter ratio 3.32 dB higher than the OFDM signal. However, in environments with higher noise levels and more challenging sea conditions, further optimization of the filters may be required to enhance target resolution and clutter handling.

Acknowledgements This work was supported by the HuiYanXingDong (62402010206).

References

1. Navya, R., Devaraju, R.: A systematic survey on radar target detection techniques in sea clutter background. In: Satapathy, S.C., Bhateja, V., Favorskaya, M.N., Adilakshmi, T. (eds.) Smart Computing Techniques and applications. Smart Innovation, Systems and Technologies, vol. 224. Springer, Singapore (2021)
2. Guidoum, N., Soltani, F., Mezache, A.: Modeling of high-resolution radar sea clutter using two approximations of the Weibull plus thermal noise distribution. Arab. J. Sci. Eng. **47**, 14957–14967 (2022)
3. Farshchian, M., Cowen, B., Selesnick, I.: Phenomenology based decomposition of sea clutter with a secondary target classifier. In: 2023 IEEE Radar Conference (RadarConf23), San Antonio, TX, USA, pp. 1–6 (2023)
4. Pei, J., et al.: A sea clutter suppression method based on machine learning approach for marine surveillance radar. IEEE J. Sel. Top. Appl. Earth Obs. Remote. Sens. **15**, 3120–3130 (2022)
5. Mou, X., Chen, X., Guan, J., Dong, Y., Liu, N.: Sea clutter suppression for radar PPI images based on SCS-GAN. IEEE Geosci. Remote. Sens. Lett. **18**(11), 1886–1890 (2021)
6. Xiao, B., Qu, W., Qiu, L., et al., OFDM integrated waveform design for joint radar and communication. Wirel. Netw. **30**(6), 5425–5436 (2023)
7. Mirabella, M., Viesti, P.D., Davoli, A., Vitetta, G.M.: Deterministic signal processing techniques for OFDM-based radar sensing: an overview. IEEE Access **11**, 68872–68889 (2023)
8. Li, J., Deng, Q., Ou, J., Wang, W.: Research on multi-carrier phase code signal against interrupted sampling repeater jamming based on MIMO radar. J. Signal Process. **38**(7), 1488–1496 (2022)

MCPA-DETR: Improving DETR with Modulated Constraint and Progressive Assignment for Accurate and Efficient Object Detection

Chuang Zhang, Yan Gui, Zuwang Pan, and Ruojun Guo

Abstract The recently developed DEtection TRansformer (DETR) establishes a new paradigm of object detection with a set of learnable queries. However, this paradigm still lags in performance behind CNN-based detectors. The limited performance is largely attributed to redundant predictions among queries and the supervision sparsity on predictions. In this paper, we present a novel MCPA-DETR with the corresponding two contributions. Firstly, we propose a query-aware region constraint to make the responsibility region of each query more focused. Secondly, we introduce a progressive label assignment to provide richer supervision signals, promoting the convergence speed. Our MCPA-DETR is efficient due to without introducing extra parameters and computational overhead. Extensive experiments validate the effectiveness of the proposed method, and our method can generalize well across existing DETR-based models. On the challenging MS COCO benchmark, our MCPA-DETR outperforms the existing models by a large margin.

Keywords Object detection · DETR · Query-aware region constraint · Progressive label assignment

C. Zhang · Y. Gui (✉) · Z. Pan · R. Guo
School of Computer Science and Technology, Changsha University of Science and Technology, Changsha, Hunan, China
e-mail: guiyan@csust.edu.cn

Engineering Research Center of Catastrophic Prophylaxis and Treatment of Road and Traffic Safety of Ministry of Education, Changsha University of Science and Technology, Changsha, Hunan, China

C. Zhang
e-mail: zhangchuang@stu.csust.edu.cn

Z. Pan
e-mail: panzuwang@stu.csust.edu.cn

R. Guo
e-mail: guoruojun@stu.csust.edu.cn

J. S. Park et al. (eds.), *Advanced Multimedia and Ubiquitous Engineering*, Lecture Notes in Electrical Engineering 1475, https://doi.org/10.1007/978-981-95-1565-3_4

1 Introduction

Object detection is one of the most fundamental computer vision tasks, aiming to locate as well as classify instances of objects in an image. It has been widely used in real-world applications, such as autonomous driving and security monitoring. Recently, Carion et al. [1] proposed a new paradigm that utilizes Transformer [2] for object detection, called DEtection TRansformer or DETR. It treats object detection as a direct set prediction problem—namely, that given a fixed small set of learnable object queries, it predicts all objects at once. Additionally, DETR simplifies the detection pipeline, eliminates the hand-designed components, and enables end-to-end detection.

Despite the impressive results achieved, DETR has certain limitations of slow convergence and inferior performance. Many follow-up works [3–12] have attempted to tackle these issues, like introducing query-specific priors [3–7]. For example, Anchor DETR [5] formulates queries as 2D anchor points, where each query only focuses on a certain region near the corresponding anchor point. DAB-DETR [6] extends 2D anchor points to 4D anchor boxes and dynamically updates them layer by layer. However, using general queries in DETR-like detectors [1, 3–12], the Transformer attention module does not well attend to the spatial ranges containing the object extremities or regions inside the object box, so that the prediction box of each query deviates from the corresponding spatial range. When the number of queries near the object is insufficient, more queries far away from the object are needed to localize the extremities, which results in repeated predictions of the same object by the queries.

Label assignment plays a crucial role in object detection, which assigns one prediction or multiple predictions to each ground-truth. DETR and its variants [3–10] employ the Hungarian matching to perform one-to-one assignments based on the global minimum matching costs, but too few queries assigned as positive examples lead to sparse supervision, significantly reducing the training efficiency. Some recent studies [11, 12] have proposed one-to-many assignment strategies to provide adequate supervision. For instance, DELA-DETR [11] introduces a label augmentation strategy to repeat the labels of ground-truth objects. Group DETR [12] uses multiple groups of object queries and performs a one-to-one assignment on each group. However, DELA-DETR conducts a non-dynamic one-to-many assignment, while Group DETR uses parallel query groups with 3300 queries and matches them with ground-truths in a one-to-many manner, which leads to an increased computational complexity.

To alleviate the above issues, we propose a novel method, namely MCPA-DETR. Firstly, we present a query-aware region constraint to narrow the spatial management range of each query and avoid the ambiguous spatial prior among queries. Subsequently, we design an efficient dynamic one-to-many matching strategy without introducing more queries, called progressive label assignment. It enables dynamic label assignment at each decoder layer and places particular training emphasis on the later decoder layers, gradually providing adequate supervision at each decoder layer.

Finally, we validate the effectiveness of our method with thorough ablation studies. Equipped with ResNet-50 [13] and ResNet-101 backbones and a DAB-DETR [6] baseline, our MCPA-DETR respectively achieves 2.4 AP and 2.1 AP gains under the training with 50 epochs.

2 Method

Our proposed MCPA-DETR is based on the basic architecture of DAB-DETR with two main changes. Figure 1 illustrates the overall framework of our method. Firstly, the query-aware region constraint is used to narrow down the spatial range of each query to the object area, which helps to produce the predictions from the corresponding object queries that well adapt to objects with various scales and aspect ratios. Secondly, we perform a progressive one-to-many label assignment between the predictions and ground-truth objects to provide richer supervision signals for training, thereby accelerating the model convergence. Next, we describe in detail our MCPA-DETR.

2.1 *Query-Aware Region Constraint*

We present position and aspect ratio similarity constraints on the queries, to make their respective areas of responsibility more focused and avoid unnecessary prediction conflicts among queries, as shown in the left part of Fig. 2. Specifically, we introduce a position constraint loss $\mathcal{L}_{\text{pos}}$ to the Hungarian loss to bound the predictions of the queries to a specific area, providing them with improved spatial position prior. In addition, we introduce an aspect ratio similarity constraint loss $\mathcal{L}_{\text{ars}}$, such that the queries are adjusted to accommodate the objects with excessive scale variance. Let B_i be the prediction box of the i-th query and A_i be the i-th anchor box. The query-aware region constraint loss $\mathcal{L}_{\text{qarc}}$ is a combination of the $\mathcal{L}_{\text{pos}}$ and $\mathcal{L}_{\text{ars}}$:

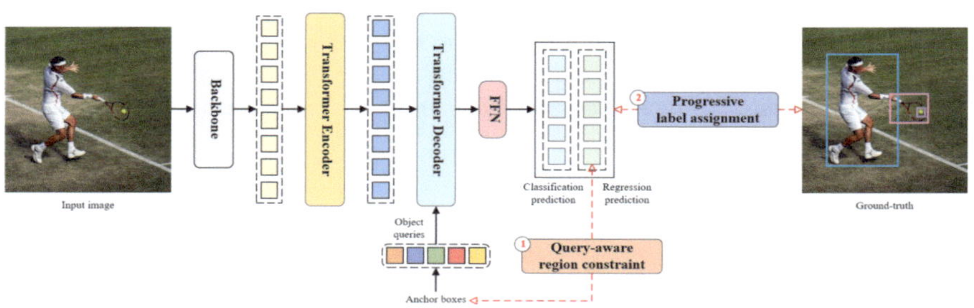

Fig. 1 The framework of the proposed MCPA-DETR

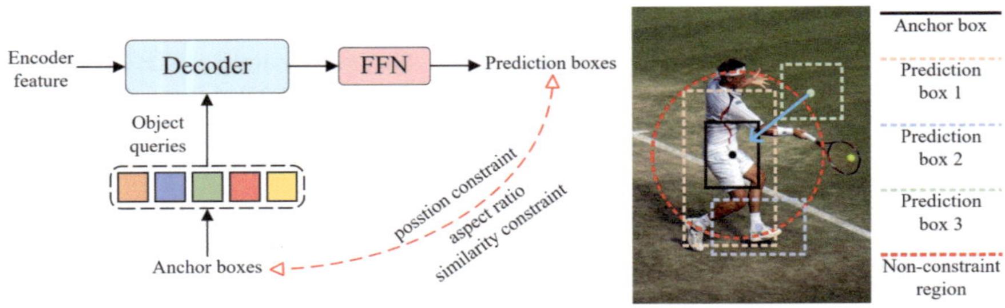

Fig. 2 Illustration for the query-aware region constraint

$$\begin{aligned}\mathcal{L}_{\text{qarc}} &= \lambda_{\text{pos}}\mathcal{L}_{\text{pos}} + \lambda_{\text{ars}}\mathcal{L}_{\text{ars}} \\ &= \frac{\lambda_{\text{pos}}}{\varepsilon_1}\sum_{i=1}^{N} 1_{\{\rho>\eta_{\text{pos}}\}}\rho_i + \frac{\lambda_{\text{ars}}}{\varepsilon_2}\sum_{i=1}^{N} 1_{\{\upsilon>\eta_{\text{ars}}\}}\nu_i\end{aligned} \tag{1}$$

$$\rho_i = \|B_i^{\{x,y\}} - A_i^{\{x,y\}}\|_2 \tag{2}$$

$$\nu_i = \frac{4}{\pi^2}\left(\arctan\frac{w_{B_i}}{h_{B_i}} - \arctan\frac{w_{A_i}}{h_{A_i}}\right)^2 \tag{3}$$

where N is the number of predicted boxes. ρ_i represents the distance between the center points of B_i and A_i. ν_i is the aspect ratio similarity between B_i and A_i. w_{B_i} (w_{A_i}) and h_{B_i} (h_{A_i}) are the width and height of B_i (A_i), respectively. η_{pos} is the distance threshold and η_{ars} is the aspect ratio similarity threshold. When the distance ρ_i exceeds the threshold η_{pos} or the aspect ratio similarity ν_i exceeds the threshold η_{ars}, penalties are imposed on B_i. ε_1 and ε_2 are the number of bounding boxes to be penalized, respectively. λ_{pos} and λ_{ars} are loss balance coefficients.

As shown in the right part of Fig. 2, for the anchor box (the black box), the model produces a prediction box 1 and a prediction box 2 (the orange and blue boxes), whose center points are inside the non-constrained region highlighted by the red dotted circle, as well as prediction box 3 (the green box) whose center point is outside the non-constrained region. In this case, the prediction box 3 is penalized, in order that the objects within the corresponding management area will be focused by the query. As a result, the queries are collaboratively in charge of their corresponding foreground objects. This avoids providing the ambiguous spatial priors to the queries.

2.2 *Progressive Label Assignment*

We propose a progressive label assignment to alleviate the sparse supervision caused by the fewer positive queries. For each decoder layer, we exploit contextual information and dynamically assign K_i predicted boxes to each ground-truth box, to impose

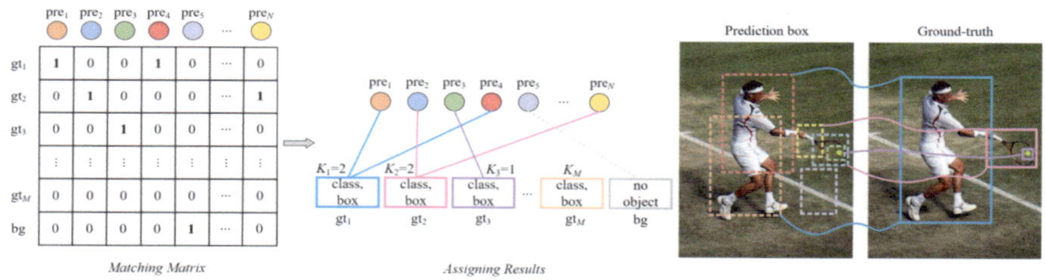

Fig. 3 Illustration for the progressive label assignment

training emphasis on the later decoder layers, as shown in Fig. 3. Specifically, for decoder layer l, we first compute the IoU scores between each ground-truth box b_i^{G} and N predicted boxes $\left\{b_j^{\mathrm{B}_l}\right\}_{j=1}^{N}$. Then, we apply a top-$k$ operation to keep the top-k entries. The number of predicted boxes assigned to each ground-truth box K_i^l can be computed as a sum of the top-k entries and rounded down. This is given by:

$$K_i^l = \left\lfloor \sum \text{top - k}\left(\left\{\text{IoU}\left(b_j^B, b_i^G\right) | j \in 1, 2, \ldots, N\right\}\right)\right\rfloor - \left\lfloor \frac{L-l}{L-3} \right\rfloor,$$
$$i = 1, 2 \ldots, M \tag{4}$$

where M is the number of ground-truth boxes. b_j^{B} is the j-th predicted box and b_i^{G} is the i-th ground-truth box. In this way, the model learns to determine the number of positive examples needed by each ground-truth based on many factors like objects' sizes, shapes, and occlusion conditions, etc.

To avoid the equivalent supervision for decoder stages, we use an adjustment factor $\left\lfloor \frac{L-l}{L-3} \right\rfloor$ to refine K_i^l layer-by-layer, where L is the number of decoder layers. This design indicates that we perform the label assignment in a progressive manner, where larger values of K_i^l are computed with the later decoder layers. Note that the minimum value of K_i^l is set to 1, so that we assign at least one predicted box to each ground-truth. The particular training emphasis on later decoder stages can efficiently correct inaccurate predictions generated during the early stages.

In addition, with the matching costs $\mathcal{C}_{\text{match}}$ between M ground-truth boxes $\left\{b_i^{\mathrm{G}}\right\}_{i=1}^{M}$ and N predicted boxes $\left\{b_j^{\mathrm{B}_l}\right\}_{j=1}^{N}$ and the number of predicted boxes assigned to the ground-truths $\left\{K_i^l\right\}_{i=1}^{M}$, we compute the matching matrix $\mathcal{H} \in \mathbb{R}^{(M+1)\times N}$, as shown in the left part of Fig. 3. In detail, for the ground-truth b_i^{G}, we use the first K_i^l predicted boxes with a smaller matching cost as the positive examples, which match the ground-truth box b_i^{G}, and record their indices $\{\sigma_i(j) | j = 1, \ldots, K_i^l\}$, so that the element in the i-th row and the $\sigma_i(j)$-th column of the matching matrix $\mathcal{H}$ is set to 1. The matched predicted boxes for background are determined similarly. If one predicted box matches multiple ground-truths, we assign the predicted box to the ground-truth object with the lowest matching cost. The final assigning result is shown in the right part of Fig. 3.

Finally, we define the one-to-many assignment $\hat{\sigma}_{\text{o2m}}$ as follows:

$$\hat{\sigma}_{\text{o2m}} = \arg\min_{\sigma_i \in C_N^M} \sum_{i=1}^{M} \sum_{j=1}^{K_i} \mathcal{C}_{\text{match}}\left(\hat{y}_{\sigma_i(j)}, y_i\right) \tag{5}$$

where C_N^M is the combination of M and N, which denotes a subset of N predicted boxes assigned to each ground-truth. $\sigma_i(j)$ are the indices of the j-th predicted boxes matched with the i-th ground-truth. $\mathcal{C}_{\text{match}}\left(\hat{y}_{\sigma_i(j)}, y_i\right)$ is the matching cost between ground-truth y_i and the predictions with indices $\sigma_i(j)$. We refer the reader to [1] for details of the matching cost.

3 Experiments

3.1 Settings

Datasets and Evaluation Metrics. We conduct all our experiments on the MS COCO [14] detection dataset and report the COCO-style Average Precision (AP).

Implementation Details. Our MCPA-DETR is built upon DAB-DETR and adopts ResNet-50 and ResNet-101 backbones. We set the initial learning rate to 1×10^{-4} and drop it at the 10-th, 20-th, 30-th, and 40-th epoch by multiplying 0.1 for the 12, 24, 36, and 50 epoch training settings. Our models are trained on 2 Nvidia 3090 GPUs with total batch size of 8. We use the query-aware region constraint loss with $\lambda_{\text{pos}} = \lambda_{\text{ars}} = 5$, $\eta_{\text{pos}} = 0.25$, and $\eta_{\text{ars}} = 0.3$. A maximum assignment number of 8 is adopted for label assignment.

3.2 Settings

Comparison with State-of-the-arts. Table 1 presents the results on the COCO validation dataset. Compared with the strong baseline DAB-DETR, our MCPA-DETR with 36 training epochs gains 1.4 AP and 1.5 AP for ResNet-50 and ResNet-101 backbones. Under 50 training epochs, our MCPA-DETR achieves 2.4 AP and 2.1 AP improvements, respectively. In addition, our MCPA-DETR is consistently better than DN-DETR [8], DELA-DETR [11] and Group DETR [12]. Meanwhile, MCPA-DETR has no additional parameters and computation compared to the baseline.

Comparison of Model Convergence. We show the convergence curves of single-scale DAB-DETR and MCPA-DETR in Fig. 4. Under the same setting, our model can

Table 1 Performance (single-scale model) comparison on the COCO validation dataset. All models use 300 queries during inference, except for DETR which uses 100 queries. Bold indicates the best results

Model	Epochs	AP	AP_{50}	AP_{75}	Params (M)	GFLOPs
DAB-DETR-R50	50	42.2	63.1	44.7	44	94
DN-DETR-R50	50	44.1	64.4	46.7	44	94
DELA-DETR-R50	50	41.9	62.6	44.8	43	91
Group DETR-R50	50	44.5	-	-	44	94
MCPA-DETR-R50 (Ours)	36	43.6	64.3	46.6	44	94
	50	**44.6**	**65.2**	**47.4**	44	94
DAB-DETR-R101	50	43.5	63.9	46.6	63	174
DN-DETR-R101	50	45.2	65.5	48.3	63	174
MCPA-DETR-R101 (Ours)	36	45.0	65.7	47.9	63	174
	50	**45.6**	**66.3**	**48.6**	63	174

outperform DAB-DETR by 3.1 AP and 2.2 AP within 24 and 36 epochs, respectively. Our MCPA-DETR with 24/36 training epochs performs better than DAB-DETR with 50 training epochs, obtaining 0.3 AP and 1.4 AP improvements.

Visualization of detection results. We also provide detection results of different methods in Fig. 5. It can be seen that our MCPA-DETR can detect the distant objects overlooked by the baseline method and there are fewer false-positive predictions.

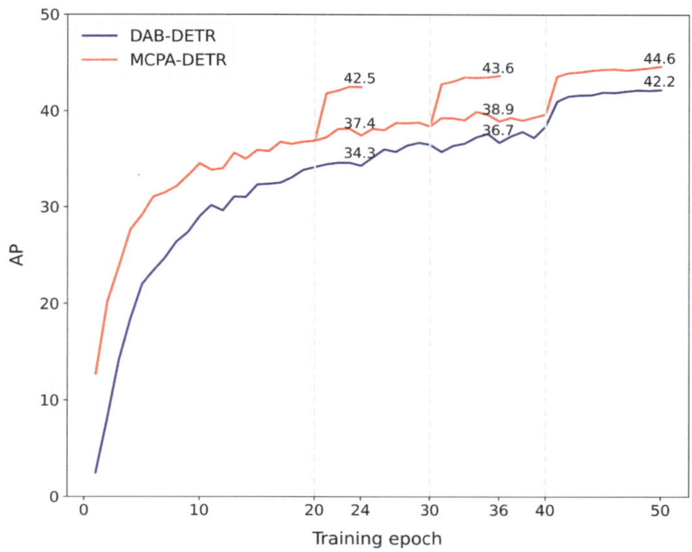

Fig. 4 Convergence curves evaluated on the COCO validation dataset for our DAB-DETR and MCPA-DETR with ResNet-50 using single-scale features

Fig. 5 Detection results of DAB-DETR and MCPA-DETR on MS COCO

Table 2 Ablations on each component in MCPA-DETR. "QARC" and "PLA" represent query-aware region constraint and progressive label assignment, respectively

Method	QARC	PLA	AP	AP_{50}	AP_{75}	AP_S	AP_M	AP_L
MCPA-DETR	√	√	**44.6**	**65.2**	**47.4**	24.3	**48.5**	**63.9**
-QARC	×	√	43.8	64.9	45.8	23.1	47.8	63.2
-PLA	√	×	43.9	64.7	46.8	**24.7**	47.8	62.7
DAB-DETR	-	-	42.2	63.1	44.7	21.5	45.7	60.3

3.3 Ablation Study

Effectiveness of Each Component. Table 2 shows the effectiveness of each component. The use of all proposed components improves the performance from 42.2 AP to 44.6 AP compared to the baseline DAB-DETR (row 5). Both QARC and PLA serve as the essential components, independently emerging 0.8 AP and 0.7 AP improvements compared to the standard model (rows 2–4). In summary, each proposed component contributes to the final performance.

4 Conclusion

In this paper, we propose MCPA-DETR for improving the detection performance by adopting two critical designs. We narrow the management range of queries with query-aware region constraint, achieving better query optimization. To address the supervision sparsity, we propose progressive label assignment to provide richer supervision signals for training. Competitive experimental results are obtained on the MS COCO benchmark and demonstrate the effectiveness of MCPA-DETR.

References

1. Carion, N., Massa, F., Synnaeve, G., Usunier, N., Kirillov, A., Zagoruyko, S.: End-to-end object detection with transformers. In: 16th European Conference on Computer Vision, vol. 12346, pp. 213–229. Springer, Glasgow (2020). LNCS
2. Vaswani, A., Shazeer, N., Parmar, N., Uszkoreit, J., Jones, L., Gomez, A.N., Kaiser, L., Polosukhin, I.: Attention is all you need. Adv. Neural Inf. Process. Syst. **30**, 5998–6008 (2017)
3. Zhu, X., Su, W., Lu, L., Li, B., Wang, X., Dai, J.: Deformable DETR: deformable transformers for end-to-end object detection. In: 9th International Conference on Learning Representations. OpenReview.net, Virtual Event (2021)
4. Meng, D., Chen, X., Fan, Z., Zeng, G., Li, H., Yuan, Y., Sun, L., Wang, J.: Conditional DETR for fast training convergence. In: IEEE International Conference on Computer Vision, pp. 3631–3640. IEEE, Montreal (2021)
5. Wang, Y., Zhang, X., Yang, T., Sun, J.: Anchor DETR: query design for transformer-based detector. In: 36th AAAI Conference on Artificial Intelligence, pp. 2567–2575. AAAI Press, Virtual Event (2022)
6. Liu, S., Li, F., Zhang, H., Yang, X., Qi, X., Su, H., Zhu, J., Zhang, L.: DAB-DETR: dynamic anchor boxes are better queries for DETR. In: 10th International Conference on Learning Representations. OpenReview.net, Virtual Event (2022)
7. Liu, Y., Zhang, Y., Wang, Y., Zhang, Y., Tian, J., Shi, Z., Fan, J., He, Z.: SAP-DETR: bridging the gap between salient points and queries-based transformer detector for fast model convergency. In: IEEE Conference on Computer Vision and Pattern Recognition, pp. 15539–15547. IEEE, Vancouver (2023)
8. Li, F., Zhang, H., Liu, S., Guo, J., Ni, L.M., Zhang, L.: DN-DETR: accelerate DETR training by introducing query denoising. In: IEEE Conference on Computer Vision and Pattern Recognition, pp. 13609–13617. IEEE, New Orleans (2022)
9. Zhang, H., Li, F., Liu, S., Zhang, L., Su, H., Zhu, J., Ni, L.M., Shum, H.: DINO: DETR with improved denoising anchor boxes for end-to-end object detection. In: 11th International Conference on Learning Representations. OpenReview.net, Kigali (2023)
10. Liu, S., Ren, T., Chen, J., Zeng, Z., Zhang, H., Li, F., Li, H., Huang, J., Su, H., Zhu, J., Zhang, L.: Detection transformer with stable matching. In: IEEE International Conference on Computer Vision, pp. 6468–6477. IEEE, Paris (2023)
11. Wang, W., Zhang, J., Cao, Y., Shen, Y., Tao, D.: Towards data-efficient detection transformers. In: 17th European Conference on Computer Vision. LNCS, vol. 13669, pp. 88–105. Springer, Tel Aviv (2022)
12. Chen, Q., Chen, X., Wang, J., Zhang, S., Yao, K., Feng, H.: Group DETR: fast training with group-wise one-to-many assignment. In: IEEE International Conference on Computer Vision, pp. 6610–6619. IEEE, Paris (2023)
13. He, K., Zhang, X., Ren, S., Sun, J.: Deep residual learning for image recognition. In: IEEE Conference on Computer Vision and Pattern Recognition, pp. 770–778. IEEE Computer Society, Las Vegas (2016)
14. Lin, T., Maire, M., Belongie, S.J., Hays, J., Perona, P., Ramanan, D., Dollar, P., Zitnick, C.L.: Microsoft COCO: common objects in context. In: 13th European Conference on Computer Vision, vol. 8693, pp. 740–755. Springer, Zurich (2014). LNCS

Robust Detection and Extraction of Lane Lines Based on Road Constraints

Zuwang Pan and Yan Gui

Abstract Challenges in lane detection, such as distractors, illumination variations and cracks, necessitate a robust detection and extraction method based on road constraints. First, a large number of images for Dongfeng Park scene was collected to create a custom lane line dataset, including line and road area annotations. Second, we define a lane line segmentation loss function with road constraints to enhance the ENet-SAD model, yielding accurate segmentation predictions. Finally, we employ the dense CRF algorithm to refine the lane line probability map. Experimental results show that our model boosts the mean IoU and F1 score by 2.23% and 2.35% compared to ENet-SAD, especially in various complex scenarios including strong light, shadow, and occlusion. The proposed loss function can further enhance the performance of existing lane line detection models, including ENet-SAD and SCNN.

Keywords Lane line detection · Deep learning · Road constraints · ENet-SAD model

1 Introduction

Environmental perception is crucial in autonomous driving, the lane detection is a hot research topic in the field of computer vision. Traditional lane detection algorithms, based on feature extraction, are prone to environmental changes, affecting their reliability. With a fast developments of deep learning, researchers are leveraging its technologies in lane detection to mitigate environmental impact.

Deep learning-based lane detection methods have yielded optimal results. Neven [1] redefined lane detection as an instance segmentation challenge. Pan [2] introduced an SCNN approach for lane line detection. Hou [3] proposed the SAD model for lane

Z. Pan · Y. Gui (✉)
School of Computer Science and Technology, Changsha University of Science and Technology, Changsha, China
e-mail: guiyan@csust.edu.cn

Z. Pan
e-mail: panzuwang@stu.csust.edu.cn

J. S. Park et al. (eds.), *Advanced Multimedia and Ubiquitous Engineering*, Lecture Notes in Electrical Engineering 1475, https://doi.org/10.1007/978-981-95-1565-3_5

detection, leveraging Knowledge Distillation. Qin [4] provided a novel definition for lane detection. Ko [5] regarded lane detection as key point detection. Recent methods [6, 7] have further enhanced lane detection accuracy. However, complex road conditions like worn lane lines and traffic congestion can compromise these methods' effectiveness.

This paper proposes a robust lane detection method based on road constraints to improve performance in scenarios with serious occlusion and distractors. Initially, videos from Dongfeng Park scenario are used to create a custom dataset by extracting images at fixed intervals and annotating lane lines and road areas. Based on the ENet-SAD model, a new network and a lane segmentation loss function with road constraints are proposed to enhance prediction accuracy. Finally, the Dense CRF algorithm is used for post-processing the probability maps of the lane lines.

2 Lane Line Dataset Construction

Dongfeng Park Scenario Lane Line Dataset. The vehicle for data collection is the autonomous driving SharingVan 1.0 + by Dongfeng Yuejoy Technology Co., Ltd., equipped with 12 cameras. Each camera captures videos with a resolution of 1920 × 1080. We mainly utilize the front-end cameras of the vehicle for data collection. The types of the road primarily are divided into straight roads, curved roads, and crossroads. The road scenarios include normal, shadows, strong light, and obstructions, with each scenario containing three lane lines. In addition to collecting lane line images, we use the image annotation tool (Labelme 4.5.7) to annotate each lane line, obtaining corresponding annotated images. The annotations for normal, shadow, strong light, and occlusion are as shown in Fig. 1.

The CULane Dataset with Road Annotation. To address the limitation that all publicly available datasets only annotate lane lines, we further annotates road areas on the CULane dataset.

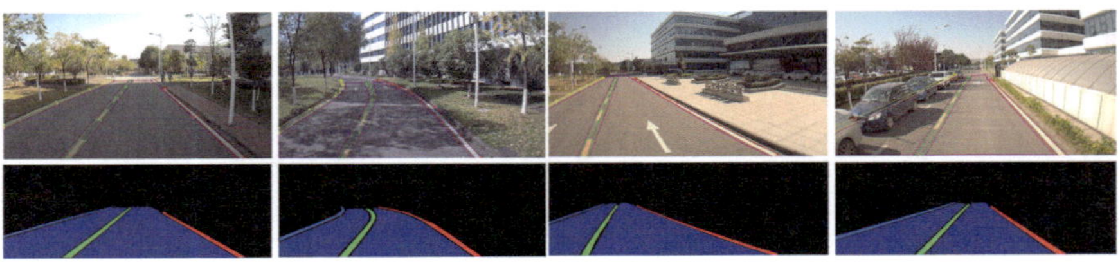

(a) Normal (b) Shadow (c) Strong Light (d) Occlusion

Fig. 1 Illustration for lane line annotations

3 Lane Detection Model

3.1 Hybrid Attention Based ENet-SAD Network

We utilize an encoder-decoder architecture, composed of four main components: an encoder, a decoder, a prediction branch network and an attention generator (AT-GEN). The overall network is depicted in Fig. 2.

The encoder consists of stages E1 to E4. E1 to E4 focus on extracting lane line features from input image. Additionally, we add a CBAM module to each of the last three stages of the encoder. The CBAM module computes spatial attention, which enables the network to focus on crucial pixel regions for classification, ignoring irrelevant areas. It also compute channel attention, which manages feature channel relationships, enhancing model performance through attention allocation across dimensions. We combine the CBAM module between E3 and E4 in a sequence and parallel manner, in order to avoid the problem of gradient vanishing and ensure network reliability.

The decoder comprises D1 and D2 stages, which use a bottleneck structure to restore spatial details and dimensions in the feature map, as shown in Fig. 2. The output of the model is a w × h × c probability map, where w and h represent the image's width and height, and c denotes the number of lane line labels. The module P1 denotes the lane line prediction branch, outputting a c-dimensional vector for lane existence estimation.

The attention generator takes the features extracted from the each encoder stage as input and produce an activation-based distillation map:

$$R^{C_m \times H_m \times W_m} \rightarrow R^{H_m \times W_m} \tag{1}$$

The attention generator is involved only during model training, ensuring no additional computational complexity during inference. Additionally, the information flow between layers allows the network to preserve scene context in deeper layers, enhancing its ability to recognize continuous, narrow lane lines.

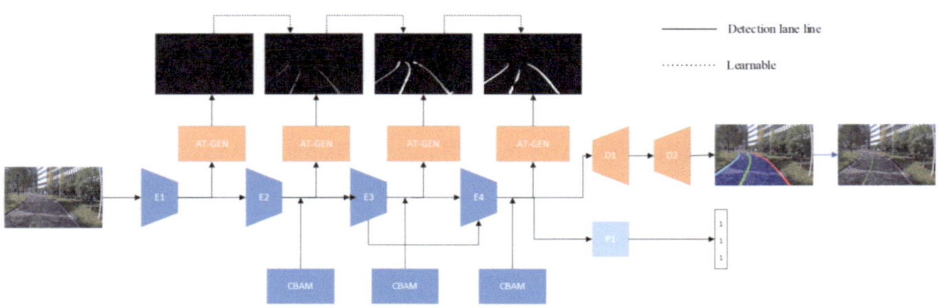

Fig. 2 Hybrid attention based ENet-SAD architecture

3.2 Roads Constrained Loss Function

We use a combination of multiple loss functions, optimizing the model by minimizing the total loss function. This is given by:

$$L_{total} = L_{seg}(s, \hat{s}) + L_{join}(s, \hat{s}) + \beta L_{exist}(b, \hat{b}) + \gamma L_{distill}(A_m + A_{m+1}) \tag{2}$$

where $L_{seg}(s, \hat{s})$ and $L_{exist}(b, \hat{b})$ are the cross-entropy loss functions, $L_{join}(s, \hat{s})$ is the segmentation loss function with joint road constraints, and $L_{distill}(s, \hat{s})$ is the distillation loss function. β and γ represent the loss weight coefficients.

The L_{join} is typically an intersection over union (IoU) loss function, which is a combination of the lane line IoU loss function and the road constraint IoU loss function:

$$L_{join}(s, \hat{s}) = \alpha_{liou} L_{liou}(s, \hat{s}) + \alpha_{riou} L_{riou}(s, \hat{s}) \tag{3}$$

where s and $\hat{s}$ are the predicted results and ground truth, respectively. α_{liou} and α_{riou} are the loss weight coefficients. L_{liou} and L_{riou} are the IoU loss functions used for the lane and road segmentations, respectively.

The distillation loss function $L_{distill}$ can distill useful contextual information to supervise the learning of network layers. By using the attention map of the previous layers to constrain the subsequent layers, it helps preserve detailed information in deep networks. It is defined as follows:

$$L_{distill}(A_m, A_{m+1}) = \sum_{m=1}^{M-1} L_d(\psi(A_m), \psi(A_{m+1})) \tag{4}$$

where L_d is the L2 loss function. A_m and A_{m+1} are the outputs of the attention generator of the m-th convolutional layer.

3.3 Lane Line Post-Processing

We further refine the probability maps using a fully connected conditional random fields (CRF) model, which formulates the refinement of the probability maps into a multi-label assignment problem, where each pixel in the image is assigned a unique category label.

4 Experiments

4.1 Settings

Implementation Details. The dataset is divided into three parts: a training set, a validation set, and a test set. We train the network on the training set and validation set. During training, the number of iterations is set to 40,000. The batch size is set to 12. The initial learning rate is set to 1e-2 and the momentum and weight decay are set to 0.9 and 2e-5 respectively.

4.2 Main Results

Comparison with State-of-the-arts. On the Dongfeng Park Scene Lane Dataset, we compare our method with LaneNet [1], ENet [8], SCNN [2], ENet-SAD [3], and enhanced ENet, SCNN, and ENet-SAD with road constraints. Figure 3 shows the segmentation results. In Figs. 3a-h, each result contains the original image with the lane segmentation, with 1–4 rows corresponding to normal, shadow, strong light, and occlusion scenarios, respectively. All compared methods produce relative poor results. As the segmentations generated by LaneNet, ENet, SCNN, and ENet-ours, detection omissions occur. LaneNet and ENet also generate segmentations with severe missing issues. This main reason is that the model perform an insufficient learning from the training samples, leading to undetected long-distance lane lines. In contrast, our method achieves the best performance by adding strong road constraints.

Table 1 reports the mean IoU and F1 score of the our method and all compared methods. The LaneNet, ENet, SCNN, and enhanced ENet methods have low mean IoU and F1 scores. For occlusion scenarios, our method achieves a mean IoU of 93.50% and a F1 score of 95.31%, surpassing the ENet-SAD-ours by over 2 points.

We also evaluate these detection methods on the CULane dataset. Figure 4 shows the experimental results. In Fig. 4a-h, the 1–4 rows represent normal, congested,

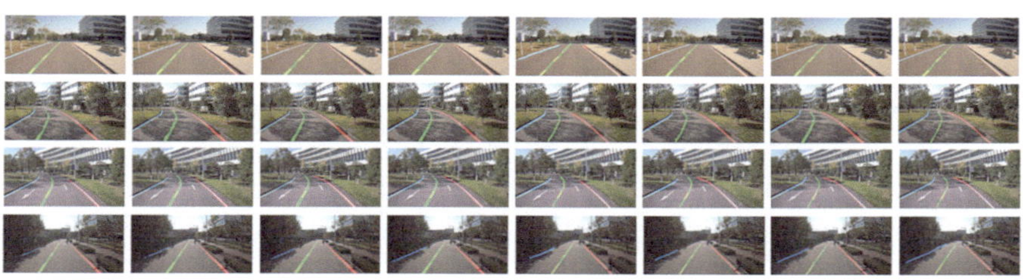

(a) (b) (c) (d) (e) (f) (g) (h)

Fig. 3 Comparison of results from different methods: **a** LaneNet **b** ENet **c** SCNN **d** ENet-SAD **e** ENet-ours **f** SCNN-ours **g** ENet-SAD-ours **h** ours

Table 1 The test results of the Dongfeng Park dataset

Methon	MIoU				MF1			
	Normal	Shadow	Strong light	Occlusion	Normal	Shadow	Strong light	Occlusion
LaneNet	92.84	89.60	91.56	87.64	94.12	91.35	93.11	88.94
ENet	92.75	90.52	92.53	88.58	94.48	92.15	94.13	90.56
SCNN	92.10	91.64	92.82	90.10	94.12	93.48	95.00	92.40
ENet-SAD	93.54	92.11	93.48	89.50	95.41	94.14	95.16	92.16
ENet-ours	92.21	91.50	92.17	90.12	94.32	93.13	94.68	92.59
SCNN-ours	93.84	93.12	93.60	90.86	96.23	94.85	95.19	92.47
Enet-SAD-ours	94.23	93.33	94.51	91.10	96.45	95.10	96.14	93.15
Ours	96.52	93.40	94.88	93.50	98.12	95.34	97.38	95.31

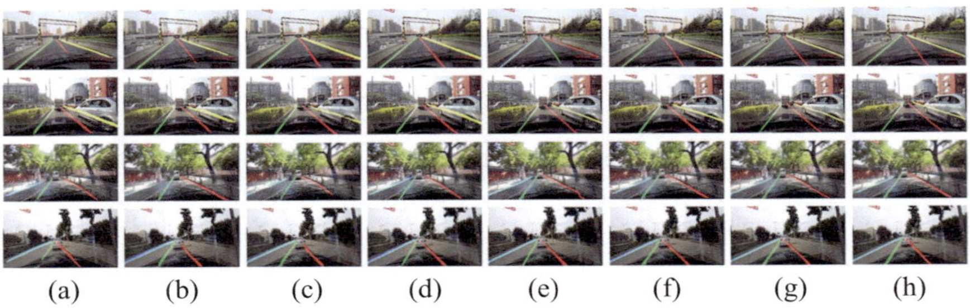

Fig. 4 Comparisons of different methods: **a** LaneNet **b** ENet **c** SCNN **d** ENet-SAD **e** ENet-ours **f** SCNN-ours **g** ENet-SAD-ours **h** ours

shadow, and no-lane scenarios, respectively. LaneNet, ENet, SCNN, and ENet-SAD perform poorly, especially in shadow and no-lane scenarios. In contrast, the enhanced methods, including ENet-ours, SCNN-ours and ENet-SAD-ours, achieve significant improvements in accuracy. Our model effectively boosts the segmentation performance.

5 Conclusion

The proposed method addresses the challenges of lane detection in scenarios where lane lines are blurry, affected by lighting conditions, or even completely occluded. The enhanced ENet-SAD model achieves a robust lane detection based on road constraints, with the use of the East Dongfeng Park dataset and the CULane dataset with lane and road annotations. Additionally, a joint road-constraint lane segmentation loss function is defined to generate accurate lane segmentation predictions. Furthermore, CRF post-processing is applied to optimize the output probability maps. Experimental results demonstrate the effectiveness of the proposed method on both the East Dongfeng Park scene lane dataset and the CULane dataset with road annotations.

Acknowledgements This work was supported by the Practical Innovation and Entrepreneurship Ability Enhancement Program for Professional Master's Degree Student at Changsha University of Science and Technology (CLSJCX22115).

References

1. Neven, D., Brabandere, B., Georgoulis, S., et al.: Towards end-to-end lane detection: an instance segmentation approach. In: 29th IEEE Intelligent Vehicles Symposium, pp. 286–291. IEEE, ChangShu (2018)
2. Pan, X., Shi, J., Luo, P., et al.: Spatial as deep: spatial CNN for traffic scene understanding. In: 32th AAAI Conference on Artificial Intelligence. AAAI Press, New Orleans (2018)
3. Hou, Y., Ma, Z., Liu, C., et al.: Learning lightweight lane detection CNNs by self attention distillation. In: 17th IEEE International Conference on Computer Vision, pp. 1013–1021. IEEE, Seoul (2019)
4. Qin, Z., Wang, H., Li, X.: Ultra fast structure-aware deep lane detection. In: 16th European Conference on Computer Vision, vol. 24, pp. 276–291. Springer, Online (2020). LNCS
5. Ko, Y., Lee, Y., Azam, S., et al.: Key points estimation and point instance segmentation approach for lane detection. IEEE Trans. Intell. Transp. Syst. **23**(7), 8949–8958 (2022)
6. Yan, L., Hu, S., Zhang, C.: Lane detection based on improved FCN. In: 33rd Chinese Control and Decision Conference, pp. 887–892. IEEE, KunMing (2021)
7. Yang, J., Zhang L., Lu H.: Lane detection with versatile atrousformer and local semantic guidance. Pattern Recognit. **133**, 109053 (2023)
8. Paszke, A., Chaurasia, A., Kim, S., et al.: ENet: A Deep Neural Network Architecture for Real-Time Semantic Segmentation, p. 02147 (2016). arXiv preprint arXiv(1606)

An Image Inpainting Method Based on Bidirectional Feature Enhancement and Multi-Scale Feature Aggregation

Ziqi Zhou, Dengyong Zhang, Rongrong Gong, Jiaxin Chen, and Fayez Alqahtani

Abstract In the process of image restoration, in order to realize the interaction between structure and texture information in the process of structural reconstruction and enhance the semantic authenticity of the restored image, on the basis of the original dual-stream generation network, we improve a dual-stream structure image recovery network based on multi-scale feature fusion algorithm. The network realizes the information interaction between structural features and texture features in space and channel in the way of feature enhancement, realizes the effective use of texture and structural information, and is conducive to generating more realistic images and preserve more semantic information. A multi-scale feature fusion network with concatenate structure is constructed to achieve the global consistency of the fusion image. By comparing experiments with other restoration networks on the Paris Street View, CelebA and Places2 datasets, it is proved that the improved image inpainting method has better objective evaluation indexes, more effective repair of the structure and texture information of damaged images, and better image inpainting performance.

Z. Zhou · D. Zhang (✉) · J. Chen
School of Computer Science and Technology, Changsha University of Science and Technology, Changsha 410114, China
e-mail: zhdy@csust.edu.cn; zhdy0@163.com

Z. Zhou
e-mail: hyzhouziqi@126.com

J. Chen
e-mail: jxchen@csust.edu.cn

R. Gong
Software College, Changsha Social Work College, Changsha 410004, China

F. Alqahtani
Software Engineering Department, College of Computer and Information Sciences, King Saud University, Riyadh 12372, Saudi Arabia
e-mail: fhalqahtani@ksu.edu.sa

J. S. Park et al. (eds.), *Advanced Multimedia and Ubiquitous Engineering*, Lecture Notes in Electrical Engineering 1475, https://doi.org/10.1007/978-981-95-1565-3_6

Keywords Image inpainting · Dual-flow structure · Multi-scale feature fusion · Generate adversarial networks

1 Introduction

The main task of image restoration is to fill in the missing information in the image, producing a result close to the real image. This is used in many fields of daily life, such as cultural relics protection, ancient photo restoration, film and television production, etc. With the continuous advancement of deep learning, most of the early deep image inpainting models were based on generative adversarial networks [1, 2]. To improve the performance of the adversarial model, Yu et al. [3] proposed a coarse-to-fine Contextual Attention framework, which improved the repair performance, but there was distortion. Subsequently, focusing on structural information extraction, the researchers proposed a series of multi-stage methods that continuously combine additional structural features [4, 5], but they were sensitive to structural accuracy. Nazeri et al. [6] proposed a texture encoder with a tandem structure, but it was unstable. In order to enable the network to learn a wider range of feature information, Liu et al. [7] proposed a cross-encoding-decoder image restoration network that mixed structure and texture. Zeng et al. [8] used dilated convolution in generative adversarial networks. These adversarial networks focused on generating more reasonable texture information and structural information, but ignored the interaction of structural texture information in the process of image reconstruction. In order to solve the above problems, inspired by the Conditional Texture and Structure Dual Generation for image inpainting (CTSDG) [9], we construct a bidirectional feature enhancement and dense connection expansion convolution module based on the adversarial network, as shown in Fig. 1. We introduce a feature enhancement module into the generator to capture global structure information while focusing on local texture information. Then, the Contextual Feature Aggregation module uses dense joins to learn higher-level semantic information from low-level details, reducing the risk of overfitting and making the most of each sample feature. We conduct multi-set experiments on three datasets: Paris Street View [10], CelebA [11], and Places2 [12], combined with the irregular mask dataset. Under the comparison of objective evaluation indicators, our feature enhancement and multi-scale feature aggregation blocks can produce good experimental results in the process of confrontation repair.

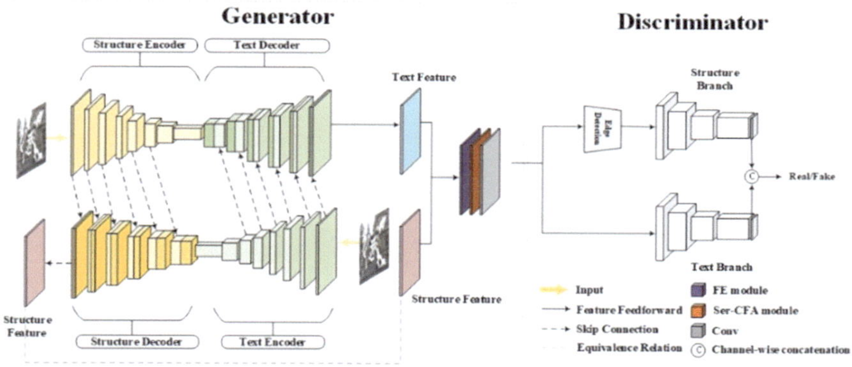

Fig. 1 Backbone model. The whole model is based on the adversarial generative network architecture, and we introduce structural and texture feature enhancement blocks (purple blocks) and feature aggregation blocks (orange blocks) to fuse the features of the double-branch output to obtain the final result.

2 Method

2.1 *Structure and Texture Bidirectional Feature Enhancement Module*

The module adopts a cross-attention mechanism to strengthen the precise matching between structural features and texture features, the local similarity between different features and the global correlation between enhanced features can be suppressed through the feature enhancement module, and the structure of the module is shown in Fig. 2a.

The input feature maps are represented as $\left(F^{\text{str}}, F^{\text{text}}\right)$, the structure feature here is $F^{\text{str}} \in R^{C\times H\times W}$ and the texture feature is $F^{\text{text}} \in R^{C\times H\times W}$. Firstly, the feature maps K and Q are obtained by convolution for F^{str} and F^{text}, where $K, Q \in R^{C_k\times H\times W}$ and C_k are the number of channels in the low-dimensional mapping space. In order to realize the cross-image information exchange, the feature representation is enhanced in the form of two branches. K and Q are transposed, and the matrices are multiplied and then transposed, and the obtained feature maps are calculated in the Softmax layer for spatial attention figures A^q and A^s of the structure branch and texture branch:

$$A^q_{ji} = \frac{\exp\left(Q_i \cdot K_j\right)}{\sum_{i=1}^{N} \exp\left(Q_i \cdot K_j\right)}, A^s = transport\left(A^q\right) \tag{1}$$

where A^q_{ij} represents the weight generated by the interaction between the ith position in the texture feature and the jth position in the structure feature.

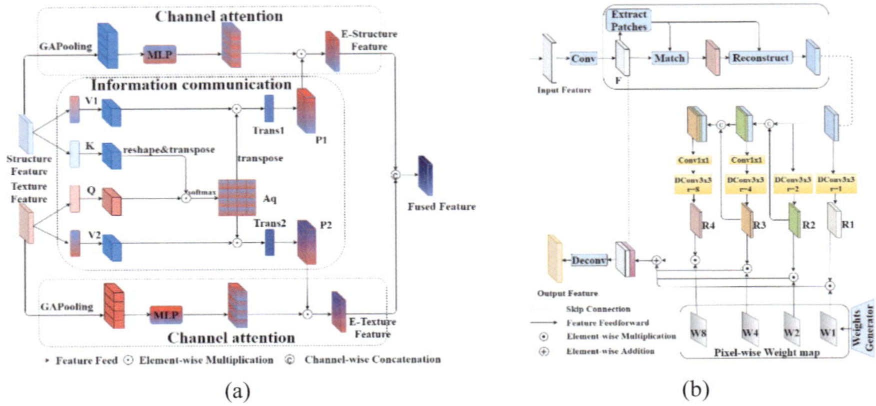

Fig. 2 Feature fusion blocks, **a** represents the structure and texture bidirectional feature enhancement module, and **b** represents the feature aggregation module of multi-scale dense connection

After the information exchange between images is completed, the In-Image Channel Attention Module is used to further enhance the structural features and texture features. The texture and structure feature maps output by the double branch are successively passed through global average pooling and multi-layer perceptron to obtain the channel attention weights, and the weights are applied to the texture structure feature map after spatial attention interaction to obtain the final enhanced texture features.

$$E^{str} = Expand\left(MLP\left(Pooling\left(F^{str}\right)\right)\right) \odot P^{str} + F^{str} \tag{2}$$

$$E^{text} = Expand\left(MLP\left(Pooling\left(F^{text}\right)\right)\right) \odot P^{text} + F^{text} \tag{3}$$

Finally, the enhanced structural feature E^{str} and texture feature E^{text} are connected in series one channel by one to obtain the integrated feature figure F_b.

2.2 Multi-Scale Densely Connected Feature Aggregation Module

In order to accurately locate and rationally utilize the effective pixels, we aggregate the enhanced structure and texture features in a multi-scale dense connection mode, which not only ensures the global consistency of the image, but also enhances the correlation between local pixels in the image, as shown in Fig. 2b.

Firstly, the input feature map F is sharded at 3×3 size to obtain a set of segmented blocks $(f_1, f_2, f_3, \cdots)$, and the cosine similarity between the blocks is calculated. The

Softmax function is then applied to calculate the attention score for each block based on the similarity between the blocks.

According to the calculated attention scores between block groups, the feature map F_{rec} is reconstructed, and the constructed feature map is combined through the extended convolutional network designed in this paper to obtain more semantic information at multiple scales. The inflated convolutional network of channel splicing superimposes and reuses the original receptive field with the receptive fields of other three scales through a densely connected pyramid structure, which effectively alleviates the problem of network degradation and achieves a higher receptive field under the condition of low expansion rate.

$$R_1 = DConv_{3\times3}^{[1]}(F_{rec}) \tag{4}$$

$$R_2 = DConv_{3\times3}^{[2]}(F_{rec}) \tag{5}$$

$$R_3 = DConv_{3\times3}^{[4]}(Conv_{1\times1}(R_2 \odot F_{rec})) \tag{6}$$

$$R_4 = DConv_{3\times3}^{[8]}(Conv_{1\times1}(R_2 \odot R_3 \odot F_{rec})) \tag{7}$$

where $DConv_{3\times3}^{[k]}$ is the deformable convolution of the convolution kernel 3×3, k is the expansion rate and $k \in \{1, 2, 4, 8\}$.

Then, the pixel-level weight feature map is used to weight the semantic features at multiple scales to obtain the final feature F_r.

3 Experiment

The model is trained on an Ubuntu 18.04 system using a single NVIDIA A30 Tensor Core GPU (24 GB), implemented using the PyTorch framework, the batch size is set to 4, the Adam optimizer is selected for parameter optimization, and we use a learning rate of 1×10^{-4} for initial training.

To verify the effectiveness of the model, we compare the model with three representative methods in the field of image inpainting, and compare the following restoration methods: partial convolution (PC) [13], cyclic feature inference (RFR) [14], and double U-Net texture structure generation (CTSDG) [9].

In Table 1, we present the results of quantitative comparison of several experimental methods on three datasets. Our model is higher than the three models compared in both PSNR and SSIM. And visually, compared with other models, the restoration results of our model are structurally more complete, and the texture details are clearer, as shown in Fig. 3.

Table 1 Our model quantitatively compares experimental results on three public datasets

	Metrics	PSNR↑				SSIM↑			
Datasets	Mask Ratio	10–20	20–30	30–40	40–50	10–20	20–30	30–40	40–50
Paris Street View	Pconv	31.56	28.33	26.11	24.12	0.940	0.893	0.838	0.772
	RFR	31.60	28.39	26.15	24.17	0.941	0.892	0.838	0.772
	CTSDG	32.73	29.30	26.89	24.81	0.950	0.906	0.854	0.790
	Ours	**33.49**	**29.92**	**27.53**	**25.30**	**0.955**	**0.914**	**0.864**	**0.803**
Celeba	Pconv	32.16	28.96	26.54	24.55	0.955	0.920	0.881	0.838
	RFR	33.01	29.41	26.73	24.62	0.959	0.922	0.880	0.832
	CTSDG	34.34	30.38	27.52	25.28	0.967	0.935	0.895	0.852
	Ours	**35.15**	**31.05**	**28.13**	**25.86**	**0.970**	**0.939**	**0.910**	**0.861**
Places2	Pconv	27.25	24.09	21.96	20.28	0.922	0.861	0.797	0.727
	RFR	27.63	24.50	22.36	20.64	0.915	0.848	0.779	0.704
	CTSDG	28.92	25.51	23.19	21.37	0.937	0.887	0.830	0.766
	Ours	**29.24**	**25.74**	**23.39**	**21.56**	**0.938**	**0.889**	**0.832**	**0.770**

The bold shows the experimental results of the proposed approach

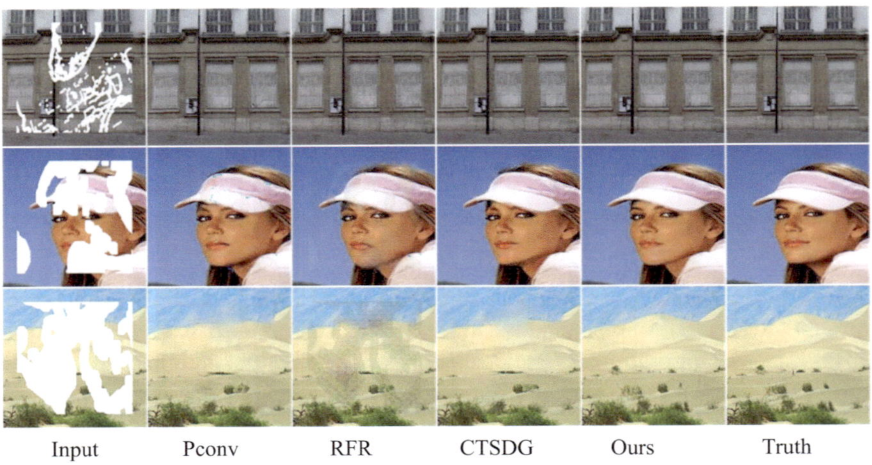

Fig. 3 Diagram of the repair results of our model. From top to bottom, the experimental results are compared on the Paris Street View, Celeba, and Places2 datasets.

4 Conclusion

In order to effectively and fully utilize the structure information and texture information in the process of image inpainting and enhance the global consistency of the restored image, we design a two-way enhancement module for texture and structure, and a feature aggregation module with multi-scale dense connection based on the original dual-stream network, and verify their effectiveness through experiments. Experimental results show that based on the training results of three datasets, the proposed model not only has higher objective evaluation indexes in image inpainting results, but also can achieve better subjective visual effects.

References

1. Goodfellow, I., Pouget-Abadie, J., Mirza, M., Xu, B., Warde-Farley, D., Ozair, S., Courville, A., Bengio, Y.: Generative adversarial nets. Adv. Neural Inf. Process. Syst. **27** (2014)
2. Pathak, D., Krahenbuhl, P., Donahue, J., Darrell, T., Efros, A.A.: Context encoders: feature learning by inpainting. In: Proceedings of the IEEE Conference on Computer Vision and Pattern Recognition, pp. 2536–2544 (2016)
3. Yu, J., Lin, Z., Yang, J., Shen, X., Lu, X., Huang, T.S.: Generative image inpainting with contextual attention. In: Proceedings of the IEEE Conference on Computer Vision and Pattern Recognition, pp. 5505–5514 (2018)
4. Xiong, W., Yu, J., Lin, Z., Yang, J., Lu, X., Barnes, C., Luo, J.: Foreground-aware image inpainting. In: Proceedings of the IEEE/CVF Conference on Computer Vision and Pattern Recognition, pp. 5840–5848 (2019)
5. Ren, Y., Yu, X., Zhang, R., Li, T. H., Liu, S., Li, G.: Structureflow: image inpainting via structure-aware appearance flow. In Proceedings of the IEEE/CVF International Conference on Computer Vision, pp. 181–190 (2019)
6. Nazeri, K., Ng, E., Joseph, T., Qureshi, F., Ebrahimi, M.: Edgeconnect: structure guided image inpainting using edge prediction. In: Proceedings of the IEEE/CVF International Conference on Computer Vision Workshops, pp. 0–0 (2019)
7. Liu, H., Jiang, B., Song, Y., Huang, W., Yang, C.: Rethinking image inpainting via a mutual encoder-decoder with feature equalizations. In: Computer Vision–ECCV 2020: 16th European Conference, Glasgow, 23–28 August 2020, Proceedings, Part II 16, pp. 725–741. Springer International Publishing, UK (2020)
8. Zeng, Y., Fu, J., Chao, H., Guo, B.: Aggregated contextual transformations for high-resolution image inpainting. IEEE Trans. Vis. Comput. Graph. **29**(7), 3266–3280 (2023)
9. Guo, X., Yang, H., Huang, D.: Image inpainting via conditional texture and structure dual generation. In: Proceedings of the IEEE/CVF International Conference on Computer Vision, pp. 14134–14143 (2021)
10. Doersch, C., Singh, S., Gupta, A., Sivic, J., Efros, A.A.: What makes paris look like paris. Commun. ACM **58**(12), 103–110 (2015)
11. Liu, Z., Luo, P., Wang, X., Tang, X.: Deep learning face attributes in the wild. In: Proceedings of the IEEE International Conference on Computer Vision, pp. 3730–3738 (2015)
12. Zhou, B., Lapedriza, A., Khosla, A., Oliva, A., Torralba, A.: Places: a 10 million image database for scene recognition. IEEE Trans. Pattern Anal. Mach. Intell. **40**(6), 1452–1464 (2017)

13. Liu, G., Reda, F.A., Shih, K.J., Wang, T.C., Tao, A., Catanzaro, B.: Image inpainting for irregular holes using partial convolutions. In: Proceedings of the European Conference on Computer Vision (ECCV), pp. 85–100 (2018)
14. Li, J., Wang, N., Zhang, L., Du, B., Tao, D.: Recurrent feature reasoning for image inpainting. In: Proceedings of the IEEE/CVF Conference on Computer Vision and Pattern Recognition, pp. 7760–7768 (2020)

Multi-Time Window Runoff Prediction for Inland River Basins in China Based on Self-Attention-LSTM Model

Jingxian Jiang, Haowei Huang, and Jin Zhang

Abstract Runoff prediction is an important and challenging nonlinear time series problem in hydrology. It is difficult to capture complex local dynamic changes and uncertainties due to nonlinear and uncertain hydrological data. Therefore, a multi-time window runoff prediction method based on Self-Attention-LSTM model is proposed. By introducing Self-Attention mechanism to dynamically learn and highlight important information at different time scales, convolutional neural network is used to extract meteorological and hydrological characteristics at different time points, and key information and LSTM output are integrated. The features are mapped to future multi-step runoff sequences to improve the model's ability to understand and predict runoff data. In this experiment, meteorological and hydrological characteristics and runoff data of Qingshuijiang River Basin in Guizhou were selected from a dataset driven by regional meteorological elements in China, and the runoff forecast in the next 3 h (one step), 1 day (8 steps) and 3 days (24 steps) was taken as the target. The experimental results showed that, compared with the classical LSTM and Conv-LSTM models, the efficiency coefficient (NSE) of Self-Attention-LSTM in runoff prediction is significantly improved. In 3 h prediction, the NSE increases by 8.2% and 7.6% respectively. In the 1-day forecast, increases of 4.1% and 9.8%; In the 3-day forecast, the increase was 6.8% and 5.0%, and the effectiveness and superiority of the Self-Attention-LSTM model in the multi-time window runoff prediction were proved through comparative analysis experiments, which improved the utilization rate of the internal information of meteorological data, and provided beneficial exploration and new ideas for the field of runoff prediction.

Keywords Long short-term memory (LSTM) · Multi-step runoff prediction · Deep learning · Convolution neural network (CNN) · Self-attention mechanism

J. Jiang · H. Huang (✉) · J. Zhang
Changsha University of Science & Technology, Changsha, China
e-mail: huanghw@csust.edu.cn

J. S. Park et al. (eds.), *Advanced Multimedia and Ubiquitous Engineering*, Lecture Notes in Electrical Engineering 1475, https://doi.org/10.1007/978-981-95-1565-3_7

1 Introduction

In the field of hydrology, accurate runoff prediction is essential for flood risk management, water resource planning and ecosystem management [1]. Traditional physical process-based runoff prediction methods [2, 3] have a deep understanding of hydrological processes on the basis of physical principles, equations and watershed characteristics. However, it is limited by model simplification and parameterization when dealing with complex terrain, soil and rainfall changes [4]. With the rapid development of computer technology and Artificial intelligence methods, significant progress has been made in the research of runoff prediction methods [5], including Artificial Neural Network model (ANN) [6] and Support Vector Machine (SVM) [7].

Traditional recurrent neural networks (RNN) are able to take previous information into account when processing time series information, but long-term training may lead to gradient problems. To solve this problem, the method based on LSTM has been widely used in the field of hydrology and has performed well. In addition, Conv-LSTM combines the advantages of one-dimensional convolutional neural network (CNN) and LSTM, which can efficiently learn internal and inter-sequence dependencies without long-term historical information, and has obvious advantages in runoff prediction [8].

Although deep learning models have achieved good predictive power in runoff prediction, these models still have certain limitations. For example, Xiang et al. [9] proposed sequential-to-sequence model based on LSTM to estimate hourly rainfall runoff, but the Seq2Seq model has the same weight of hidden layer output vector in the transmission process, thus giving the same importance to different information transmission. The processing of different time series data information is not accurate, to better solve the problem of runoff prediction with highly nonlinear characteristics, a single model is no longer able to accurately predict, and the combination of multiple models can improve the forecasting power. Therefore, we construct a Self-Attention-LSTM model.

The main contributions of this study are as follows: First, the combination of Self-Attention mechanism and Long Short-Term Memory network (LSTM) can effectively capture the dependency relationship between different positions in the sequence. Secondly, the test is conducted based on the regional surface meteorological factor driven data set in China, and it is compared with the LSTM and Conv-LSTM benchmark models. The experimental results show that the Self-Attention-LSTM model has better performance. Third, the model can make multi-time window prediction. The forecast for the future covers different time spans, from short-term forecast to medium-term forecast, which can understand the change law of runoff from different angles, and provide multi-level and multi-angle reference for water resources management, flood control and disaster warning, which is particularly important for hydrological research.

Table 1 Variables used in the experiment

Variable	Unit
Surface downward longwave radiation	Wm^{-2}
Surface downward shortwave radiation	Wm^{-2}
Total precipitation	mm
Surface pressure	Pa
Specific humidity	$kgkg^{-1}$
Air temperature	K
Wind speed	ms^{-1}

2 Study Area and Dataset

The Self-Attention-LSTM model proposed in this paper is used to forecast runoff in Qingshuijiang River basin in China. The Basang Hydrologic observation station of Sanbanxi hydropower Station in the lower reaches of Qingshuijiang River is taken as the target station for multi-time window runoff prediction. Qingshuijiang River originates from Guiding County, Qiannan Prefecture, Guizhou Province. It is the second largest river in Guizhou province. Its inland river length is 452.2 km, its river surface width is 40–50 m, its average slope is 3.97‰, its general flood rise is 4 m, the general flood flow is 978 m^3/s, and the average multi-year flow is 355 m^3/s.

In this paper, the China Meteorological Driven Dataset (CMFD) [10] is used as the source of meteorological driven data. From the data set of CMFD [10], seven feature data of 162 grid points in the region of longitude 107.35°E to 109.05°E and latitude 26.15°N to 26.95°N, were selected as the driving factors of the model. All of the LSTM models used in this study used the seven variables listed in Table 1 as inputs.

3 Multi-Time Window Runoff Prediction Model

Based on the good performance of LSTM in studying sequence problems, For Qingshuijiang River Basin in Guizhou, in order to better cope with the challenges still faced by the model when dealing with variable length sequences and irregular time series, we further introduce the Self-Attention mechanism and propose a Self-Attention-LSTM model to better capture the dependencies of different positions in the sequence, and pay attention to each element in the sequence at the same time, so as to improve the performance and accuracy of the model.

3.1 Self-Attention-LSTM

Self-Attention-LSTM first receives the runoff feature data and uses convolution operation to extract the temporal and spatial features in the runoff feature data. Then, through the LSTM layer, the feature sequence output by the convolutional layer is modeled, and the dependencies and patterns in the sequence are captured. The detailed structure of the Self-attention-LSTM model is shown in Fig. 1 (see Fig. 1). the Self-Attention mechanism is applied to the LSTM output to calculate the attention weights between different time steps in the sequence, obtain the context information of each time step, and fuse the context information with the LSTM output to obtain richer representation capability and better time series feature expression. Finally, a shared linear unit is used to map the features of the previous layer fusion to a multi-step runoff sequence for the future. Self-Attention-LSTM uses convolutional operations, LSTM and Self-Attention mechanisms to better understand sequence data, compared with the benchmark model LSTM and Conv-LSTM, it has better performance in dealing with time series problems.

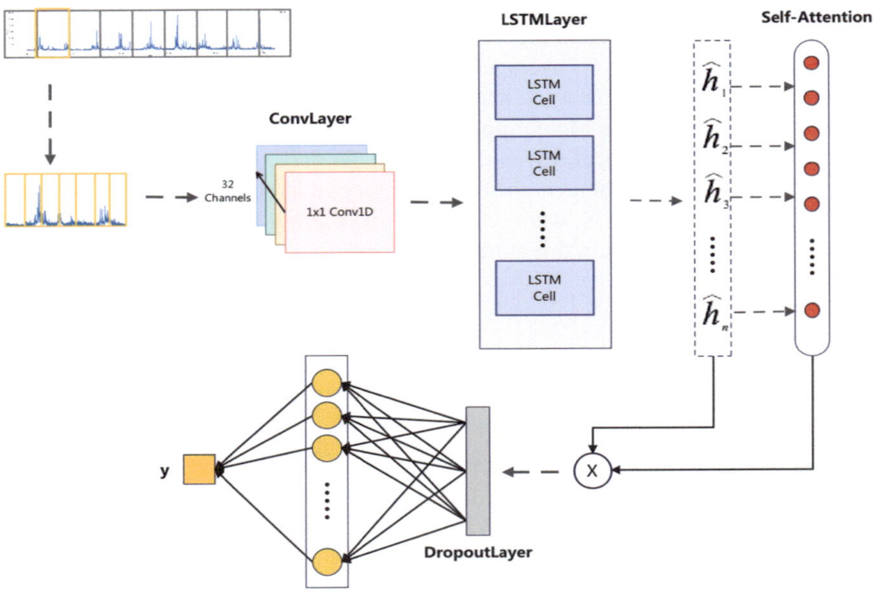

Fig. 1 Self-attention-LSTM model structure

4 Experiment Description

4.1 Benchmark Models and Experimental Process

In the process of the experiment, meteorological and hydrological characteristics and runoff data of Qingshuijiang River Basin in Guizhou were selected as the data set driven by regional meteorological elements in China, and typical LSTM [11] and Conv-LSTM [8] were selected as the benchmark model. The historical data were divided according to different time Windows, and different performance evaluation indicators were used to evaluate the performance of the model. Finally, the results are analyzed.

4.2 Evaluation Metrics

In order to comprehensively evaluate the runoff simulation performance of Self-Attention-LSTM model, we used Nash–Sutcliffe efficiency (NSE), Pearson's correlation coefficient (Pearson r) were evaluated. In addition, to emphasize the importance of extreme runoff problems and early warning of possible extreme disasters, To this end, we used the deviation values of 2% peak flow deviation (FHV) and midcourse slope deviation (FMS) divided by Yilmaz et al. [12] on the flow persistence curve to evaluate the prediction accuracy of the model for each flow segment.

$$NSE = 1 - \frac{\sum_{t=1}^{T}(O_t - M_t)^2}{\sum_{t=1}^{T}\left(O_t - \overline{O}\right)^2} \tag{1}$$

where, $t = 1,\ldots,T$ is the index of the runoff series, T is the length of the runoff series, O_t and M_t are the observed runoff and the simulated runoff value of the model respectively, $\overline{O}$ is the average observed runoff value, $NSE \in (-\infty, 1)$, The closer it is to 1 of NSE, the better the match between the simulated and observed values of the model.

$$\text{Pearson}r = \frac{\sum_{t=1}^{T}\left(M_t - \overline{O}\right)^2}{\sum_{t=1}^{T}\left(O_t - \overline{O}\right)^2} \tag{2}$$

The Pearson correlation coefficient ranges from − 1 to 1. r is an index to judge the degree of linear correlation between the measured runoff and the predicted runoff. The closer the absolute value of r is to 1, the stronger the linear correlation.

Table 2 NSE and Pearson r results of the three models in multi-time Windows

Model	3H		1D		3D	
	NSE	Pearson r	NSE	Pearson r	NSE	Pearson r
LSTM	0.76795	0.91220	0.70183	0.85039	0.66361	0.81683
Conv-LSTM	0.77255	0.89642	0.66587	0.83449	0.67503	0.82982
Self-Attention-LSTM	0.83139	0.91195	0.73121	0.84925	0.70883	0.84609

5 Results and Discussion

In order to verify the prediction ability of Self-Attention-LSTM model, this paper compared the prediction results of LSTM and Conv-LSTM model with the same input. Table 2 shows the runoff prediction results of the three models at steps 1, 8 and 24. Based on the comparison results in Table 2, it can be seen that in the 3H runoff prediction task, the NSE index of the three models is above 0.75, and the NSE of the Self-Attention-LSTM model is 0.8313, the best performance, while the Pearson r index is above 0.89. The results show that the Self-Attention-LSTM model has a very high prediction accuracy for 3-h runoff compared with the benchmark model, and the correlation between simulated runoff and actual observed runoff is very high. In the prediction of 1-day runoff, the NSE index of all models decreased, and the Self-Attention-LSTM model could also reach more than 0.73, showing a high prediction accuracy for 1D prediction. In the 3-day runoff prediction task, the NSE value of the Self-Attention-LSTM model also reached 0.7, and the Pearson r index did not decrease significantly, indicating that the model can obtain relatively high prediction accuracy in response to the multi-step runoff prediction task.

In general, the prediction effect of Self-Attention-LSTM is better than that of LSTM and Conv-LSTM, from this analysis, it can be seen that Self-Attention-LSTM model can obtain good prediction accuracy in both single-step and multi-step prediction, and can well complete the multi-time window runoff prediction task.

6 Conclusion

Multi-time window runoff prediction method based on Self-Attention-LSTM model and two benchmark models, LSTM and Conv-LSTM, were carried out several experiments in Qingshuijiang River basin and compared and analyzed. The Self-Attention-LSTM model has obvious advantages over the benchmark model in the multi-time window runoff prediction. The experimental results show that the Self-Attention-LSTM model has the best performance in the prediction of 1 step (3 h), 8 step (1 day) and 24 step (3 days) in the future, and the NSE index reaches 0.8313, 0.7312 and 0.7088, especially in the 3-h prediction task. NSE and KGE are significantly better than other models, which shows the excellent performance of the model. At the same time, the combination of physical model and Self-Attention-LSTM model should be considered in the future, and the advantages of the two methods should be fully utilized to study a better runoff prediction model based on Self-Attention.

References

1. Young, C.C., Liu, W.-C.: Prediction and modelling of rainfall–runoff during typhoon events using a physically-based and artificial neural network hybrid model. Hydrol. Sci. J. **60**(12), 2102–2116 (2014)
2. Fidal, J., Kjeldsen, T.R.: Accounting for soil moisture in rainfall runoff modelling of urban areas. J. Hydrol. **589**, 125122 (2020)
3. Huo, W., Lia, Z., Zhang, K.E., et al.: GA-PIC: an improved Green-Ampt rainfall-runoff model with a physically based infiltration distribution curve for semi-arid basins. J. Hydrol. **586**,124900 (2020)
4. Gupta, H.V., Clark, M.P., Vrugt, J.A., et al.: Towards a comprehensive assessment of model structural adequacy. Water Resour. Res. **48**, W08301 (2012)
5. Ibrahim, K.S.M.H., Huang, Y.F., Ahmed, A.N., et al.: A review of the hybrid artificial intelligence and optimization modelling of hydrological streamflow forecasting. Alex. Eng. J. **61**(1), 279–303 (2021)
6. Kumar, P.S., Praveen, T.V., Prasad, M.A.: Artificial neural network model for rainfall-runoff -a case study. Int. J. Hybrid Inf. Technol. **9**(3), 263–72 (2016)
7. Bray, M., Han, D.: Identification of support vector machines for runoff modelling. J. Hydroinformatics **6**(4), 265–280 (2004)
8. Van, S.P., Le, H.M., Thanh, D.V., et al.: Deep learning convolutional neural network in rainfall–runoff modelling. J. Hydroinf. **22**(3), 541–561 (2020)
9. Xiang, Z., Yan, J., Demir, I.: A rainfall-runoff model with LSTM based sequence-to-sequence learning. Water Resour **56**, e2019WR025326 (2019)
10. Yang, K., He, J.: China Meteorological Forcing Dataset (1979–2018) [DS] (2015)
11. Kratzert, F., Klotz, D., Brenner, C., et al.: Rainfall–runoff modelling using long short-term memory (LSTM) networks. Hydrol. Earth Syst. Sci. **22**(11), 6005–6022 (2018)
12. Yilmaz, K.K., Gupta, H.V., Wagener, T.: A process-based diagnostic approach to model evaluation: application to the NWS distributed hydrologic model. Water Resour. Res. **44**, W09417 (2008)

Bridge Component Detection Based on Improved Object Detection Algorithm YOLOV7

Peng Jian, Quanjing Zhang, and Dengyong Zhang

Abstract The accurate positioning of bridge components is helpful to the health monitoring and maintenance management of bridge structures. However, the traditional bridge inspection often requires a lot of manpower and time, and the detection efficiency and accuracy are also very low. To address this issue, this paper proposes a bridge component object detection method based on improved YOLOV7. Firstly, this paper created the Bridge Component Dataset (BC-Dataset), consisting of 1200 images, including three types of structures: bridges, piers, and girders. Secondly, in order to enhance detection accuracy in complex scenes, this paper incorporated the dynamic sparse attention (the biformer attention) module and Wise-IoU into the YOLOV7 network. Finally, experiments were conducted on the BC-Dataset, achieving a model accuracy of 91.19%. The experimental results demonstrate that our method outperforms traditional methods in terms of precision and accuracy in bridge component object detection tasks. This research is of great significance for the safety and monitoring of bridge structures, providing strong support for research and applications in related fields.

Keywords YOLOV7 · Bridge component · Biformer attention · Wise-IoU

1 Introduction

Up until the end of 2023, the total number of public bridges in China reached 1.032 million, including 8816 super large bridges. Due to the large number of bridges, distinguishing bridge components quickly and then locating and classifying them is one of the important tasks in bridge operation and maintenance phases. In practice, manual visual inspection is commonly used for maintenance purposes to assess the current condition of bridges. However, manual visual inspection is time-consuming, costly, and labor-intensive. In recent years, computer vision based on deep learning technology has been considered a promising automated tool for damage detection

P. Jian · Q. Zhang · D. Zhang (✉)
Changsha University of Science and Technology, Changsha, China
e-mail: 21208051646@stu.csust.edu.cn

J. S. Park et al. (eds.), *Advanced Multimedia and Ubiquitous Engineering*, Lecture Notes in Electrical Engineering 1475, https://doi.org/10.1007/978-981-95-1565-3_8

and classification. However, due to the fact that close-up images do not include the overall structural background, it is difficult to determine the local position of damage. Understanding the damage location and related structural components is crucial for interpreting the overall structural health. Compared to 2D image data, point clouds with additional dimensions in 3D space can be used for identifying bridge components [1]. However, the acquisition of 3D point clouds typically requires special equipment such as laser scanners or depth cameras. The use of these devices may be limited by environmental conditions, such as weather and lighting factors. Nowadays, more and more deep neural network component recognition methods [2–6] have been studied.

The model is improved based on YOLOV7 [7]. YOLOV7 is currently one of the best single-stage detection algorithms. To obtain better detection results, this paper attempted to train it on the BC-Dataset. The BC-Dataset is a dataset spent a lot of time capturing and annotating from various angles for bridges. BC-Dataset conducted extensive data statistics and benchmark tests to ensure the quality of the dataset, providing reliable training data. After experimentation, this paper found that there were still some areas for improvement in the initial version of YOLOV7—for instance, in challenging scenes or obscured areas, the algorithm often made mistakes. To address these issues, this paper combined two new methods, as shown in Fig. 1. The model integrated a dynamic sparse attention module called Biformer Routing Attention (BRA) [8]. Additionally, this paper introduced Wise-IoU [9]. Compared to the conventional IoU, Wise-IoU can more comprehensively consider the relative size, position, and similarity of targets, providing more accurate evaluation results and better adaptability in different scenarios and target attributes.

2 Related Works

2.1 BC-Dataset

Bridge images from different scenarios play a crucial role in training models. However, the lack of public datasets poses a significant challenge to AI research in the industrial sector. To address this issue, this paper have created a dataset called BC-Dataset, which includes images of bridges from various environments. It captured bridge images from different locations in Hunan Province, selected and enhanced the photos for clarity, and compiled a comprehensive dataset. As shown in Fig. 2, to showcase the condition of bridges at different times, this dataset captured images in the morning, noon, evening, and night time periods.

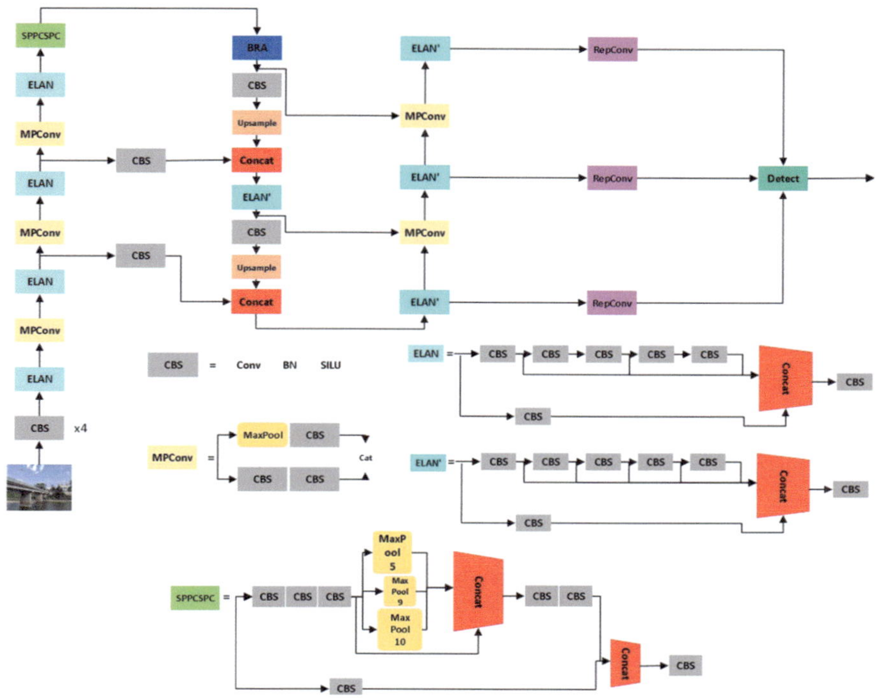

Fig. 1 Bridge component detection framework based on improved YOLOv7

Fig. 2 Bridges in different environments in BC-dataset

2.2 *Yolov7 + BRA*

Considering that YOLOv7 adopts a single neural network model and has a high inference speed in handling object detection tasks, as well as high performance in object detection tasks, accurately detecting target objects in images or videos and can handle multiple categories of objects. Therefore, this paper adopt YOLOv7 as the base model.

The process of BRA can be easily divided into three steps (as shown in Fig. 3): Firstly, assuming input a feature map, it divide it into several regions and obtain queries, keys, and values through linear mapping. Secondly, it construct a directed graph using an adjacency matrix to find the participation relationships with different

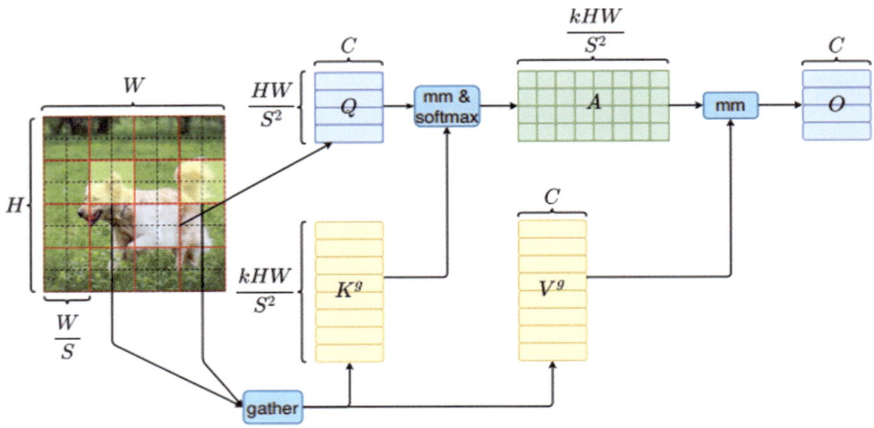

Fig. 3 Bi-level routing attention

key-value pairs, which can be understood as the regions that should participate given a certain region. Finally, with the routing index matrix from region to region, it can apply fine-grained token-to-token attention.

This paper have improved the YOLOv7-BRA (YOLOv7 with Bi-level Routing Attention) model, and the new structure is shown in Fig. 1. This paper added the BRA module at the end of the backbone network. When introducing the BRA module, This paper considered three placement options:(1) replacing all convolutional layers with convolutions with BRA;(2) placing BRA in the head network section;(3) placing BRA in the backbone network section. Choosing the first method would result in a large and difficult-to-train model, impacting the inference speed. Considering that the role of BRA is to focus on specific regions of the image rather than the entire image, this paper believe this is a crucial part of feature extraction. Therefore, this paper chose to place BRA in the backbone network section.

2.3 *YOLOV7 + WISE-IoU*

Wise-IoU provides a wise gradient gain allocation strategy. This strategy reduces the competitiveness of high-quality anchor boxes and also reduces the harmful gradients generated by low-quality examples, leading to a more accurate evaluation of object detection results, helping improve algorithms, and enhancing detection accuracy.

Wise-IoU v1 is defined as:

$$\mathcal{L}_{\mathrm{WIoUv1}} = \mathcal{R}_{\mathrm{WIoU}}\mathcal{L}_{\mathrm{IoU}}\mathrm{xc} \tag{1}$$

$$\mathcal{R}_{\mathrm{WIoU}} = \exp\left(\frac{\left(\mathrm{x} - \mathrm{x}_{\mathrm{gt}}\right)^2 + \left(\mathrm{y} - \mathrm{y}_{\mathrm{gt}}\right)^2}{\left(\mathrm{W}_{\mathrm{g}}^2 + \mathrm{H}_{\mathrm{g}}^2\right)^*}\right) \tag{2}$$

Wise-IoU v2 is defined as:

$$\mathcal{L}_{\mathrm{WIoUv2}} = \left(\frac{\mathcal{L}_{\mathrm{IoU}}^*}{\mathcal{L}_{\mathrm{IoU}}}\right)^{\gamma} \mathcal{L}_{\mathrm{WIoUv1}} \tag{3}$$

Wise-IoU v3 is defined as:

$$\mathcal{L}_{\mathrm{WIoUv3}} = \mathrm{r}\mathcal{L}_{\mathrm{WIoUv1}} \tag{4}$$

$$r = \frac{\beta}{\delta\alpha^{\beta-\delta}} \tag{5}$$

3 Experiment and Analysis

3.1 Experimental Environment and Dataset

The code is compiled and tested using Python 3.8, with the development tool being PyCharm. The core computer vision library used is python-OpenCV 4.1.5, and PyTorch v1.9.0 is employed as the deep learning framework for model training with CUDA version 11.2. The operating system is Ubuntu 18.04.6, and the CPU is an Intel(R) Xeon(R) Silver 4314 CPU with 16 CPU cores. The GPU is an NVIDIA A30 with 24GB of memory, and the system has 62GB of RAM.

The dataset used for the experiment is BC-Dataset, with a training to validation set ratio of 3:1.

3.2 Analysis of Experimental Results

In Table 1, it can be seen that mAP@0.5 represents the average precision at an IoU threshold of 0.5, while mAP@0.5:0.95 represents the average precision within the IoU threshold range of 0.5 to 0.95. Here, “W1”, “W2”, and “W3” refer to Wise-IoU versions 1, 2, and 3, respectively. This paper conducted comprehensive comparative tests on the performance of YOLOv7 with BRA integration, Wise-IoU v1, Wise-IoU v2, and Wise-IoU v3. The research results indicate that the model outperforms the original YOLOv7. By incorporating the method, the final mAP@0.5 has improved by 1.5%. Additionally, the method has increased precision and mAP@0.5:0.95 by

Table 1 Comparison of detection results among YOLOv7, YOLOv7 + BRA, and YOLOv7 + WISE-IoU

Method	Precision (%)	Recall (%)	Map50 (%)	Map50:95 (%)
YOLOV7	89.8	83.9	90.0	63.5
YOLOV7 + BRA (ours)	91.1	84.5	91.0	64.2
YOLOV7 + W1(ours)	89.7	79.3	89.5	60.1
YOLOV7 + W2(ours)	90.1	83.0	90.5	63.4
YOLOV7 + W3(ours)	91.2	**84.8**	90.9	63.5
YOLOV7 + BRA + W3(ours)	**92.3**	84.6	**91.5**	**64.5**

Table 2 Comparison of YOLOv7 + BRA + W3 with other excellent methods

Method	Precision (%)	Recall (%)	Map50 (%)	Map50:95 (%)
YOLOV8	91.9	80.5	87.3	62.7
YOLOV7	89.8	83.9	90.0	63.5
YOLOV6	79.1	70.6	83.6	56.7
YOLOV5	91.1	82.3	88.3	64.0
CST-YOLO	90.0	84.0	90.3	62.3
YOLOV7 + BRA + W3(ours)	92.3	84.6	91.5	64.5

2.5% and 1%, respectively, demonstrating the effectiveness of the research in this paper.

Table 2 compares the textual method proposed method with the latest YOLO series, where CST-YOLO is a method combining YOLOv7 with Swin Transformer. From the data in the table, it is evident that the method outperforms others in various aspects, showcasing the superiority of the algorithm in this paper.

4 Conclusion

In summary, this paper introduces a bridge component detection algorithm based on the improved YOLOv7 that accurately detects and identifies bridge-related components. This paper have created the BC-Dataset, which includes three main bridge components, filling the gap in datasets for the intersection of computer vision and bridge engineering research. By incorporating the biformer attention module and Wise-IoU into the YOLOv7 network to enhance detection accuracy, the method achieved an mAP@0.5 of 91.5%, which is a 1.5% improvement over previous results. Experimental results demonstrate that the model outperforms the original YOLOv7 in terms of precision, mAP@0.5, and mAP@0.5:0.95. The proposed method in this paper holds significant application value for the intersection of computer vision and bridge engineering research, contributing to improving recognition results. Overall, the results validate the effectiveness of the approach. The code and BC-Dataset are available at https://github.com/AutismZZ/YOLOV7-BRA-WISE_IOU.

References

1. Yulan Guo et al. "Deep learning for 3d point clouds: A survey".In: IEEE transactions on pattern analysis and machine intelligence 43.12 (2020), pp. 4338–4364.
2. Y.-J. Cha, W. Choi, O. Büyüköztürk, Deep Learning-Based Crack Damage Detection Using Convolutional Neural Networks: Deep learning-based crack damage detection using CNNs, Comput.-Aided Civ. Infrastruct. Eng. 32 (5) (2017) 361–378, https://doi.org/10.1111/mice:12263.
3. Cha, Y.J., Choi, W., Suh, G., Mahmoudkhani, S., Büyüköztürk, O.: Autonomous structural visual inspection using region-based deep learning for detecting multiple damage types. Comput.-Aided Civ. Infrastruct. Eng. **33**(9), 731–747 (2018). https://doi.org/10.1111/mice.12334
4. Xu, Y., Bao, Y., Chen, J., Zuo, W., Li, H.: Surface fatigue crack identification in steel box girder of bridges by a deep fusion convolutional neural network based on consumer-grade camera images. Struct. Health Monit. **18**(3), 653–674 (2019). https://doi.org/10.1177/1475921718764873
5. Hoskere, V., Narazaki, Y., Hoang, T.A., Spencer, B.F.: MaDnet: multi-task semantic segmentation of multiple types of structural materials and damage in images of civil infrastructure. J. Civ. Struct. Health Monit. **10**(5), 757–773 (2020). https://doi.org/10.1007/s13349-020-00409-0
6. Liang, X.: Image-based post-disaster inspection of reinforced concrete bridge systems using deep learning with Bayesian optimization. Comput.-Aided Civ. Infrastruct. Eng. **34**(5), 415–430 (2019). https://doi.org/10.1111/mice:12425
7. Wang, C.Y., Bochkovskiy, A., Liao, H.Y.M.: YOLOv7: trainable bag-of-freebies sets new state-of-the-art for real-time object detectors. arXiv (2022)
8. Zhu, L., Wang, X., Ke, Z., et al.: BiFormer: vision transformer with bi-level routing attention. arXiv preprint arXiv:2303.08810 (2023)
9. Tong, Z., Chen, Y., Xu, Z., et al.: Wise-IoU: bounding box regression loss with dynamic focusing mechanism. arXiv preprintarXiv:2301.10051 (2023)

Image Splicing Detection Based on Deformable Large Kernel Attention Network

Dengyong Zhang, Ningjing Jiang, and Feng Li

Abstract Image splicing can easily be used for illegal activities, which could have a negative impact on society. Therefore, we propose an image splicing detection algorithm based on a deformable large kernel attention network. First, a hybrid structure encoder is used to extract local and global features in the spliced image. Then a deformable large kernel attention module is designed, which we add to the hybrid structure encoder and skip connection respectively. And finally a decoder is used to reconstruct the features obtained from the hybrid structure encoder and the features received through skip connections to obtain the prediction mask at the original resolution size. A large number of experiments show that the method proposed in this article can accurately locate the splicing area, which is significantly better than other state-of-the-art methods.

Keywords Image splicing detection · Transformer · Deformable large kernel attention

D. Zhang (✉) · N. Jiang · F. Li
Hunan Provincial Key Laboratory of Intelligent Processing of Big Data on Transportation, School of Science Computer and Technology, Changsha University of Science and Technology, Changsha 410004, China
e-mail: zhdy@csust.edu.cn

School of Science Computer and Technology, Changsha University of Science and Technology, Changsha 410004, China

N. Jiang
e-mail: 21208051580@stu.csust.edu.cn

F. Li
e-mail: lif@csust.edu.cn

J. S. Park et al. (eds.), *Advanced Multimedia and Ubiquitous Engineering*, Lecture Notes in Electrical Engineering 1475, https://doi.org/10.1007/978-981-95-1565-3_9

1 Introduction

Image splicing tampering detection requires locating the splicing area in the image. Existing image splicing tampering detection methods can be roughly divided into two categories: detection methods based on traditional manual features [1–3] and detection methods based on deep learning [4–7]. Most traditional detection methods extract specific image fingerprints for detection, such as color filter array interpolation [1], sensor noise [2], illumination inconsistency [3], etc. This type of method is highly targeted and only extracts one feature of the image, which results in only one aspect of the image being considered. When the tampering method is unknown, there will be poor robustness. In recent years, deep learning theory and technology have developed rapidly and are gradually being applied in various fields. Most of the existing deep learning-based image splicing detection methods use convolutional neural networks (CNN) as feature extractors, such as RRU-Net [6], CAT-Net [7], etc. Since the receptive field of the convolution operation is limited and only operates on local areas of the image, the global dependencies between pixels cannot be modeled. Furthermore, most existing works ignore that the size of tampered regions varies, and it is difficult to locate tampered regions of different sizes.

To solve the above problems, we propose an image splicing detection algorithm based on a deformable large kernel attention network. The proposed network adopts an encoder-decoder structure, first designing a hybrid structure encoder, and simultaneously utilizing CNN and Transformer [8] to explore the local and global relationships of the image. Then a deformable large kernel attention (D-LKA) module is proposed to solve the deformation problem of features at different scales, and it is embedded into a hybrid structure encoder and skip connection to further refine the features. Finally, the decoder is used to reconstruct the features obtained from the hybrid structure encoder and the features received through skip connections to achieve pixel-level splicing area prediction.

2 Proposed Method

The image splicing detection model proposed in this paper is shown in Fig. 1. The entire network includes a hybrid structure encoder, a deformable large kernel attention module and a decoder. The input of the model is a 512×512 three-channel RGB image, and the output is a 512×512 single-channel pixel-by-pixel image mask. Where the backbone used by CNN is the first four layers of ResNet50 [9], and the backbone used by Transformer is ViT [10].

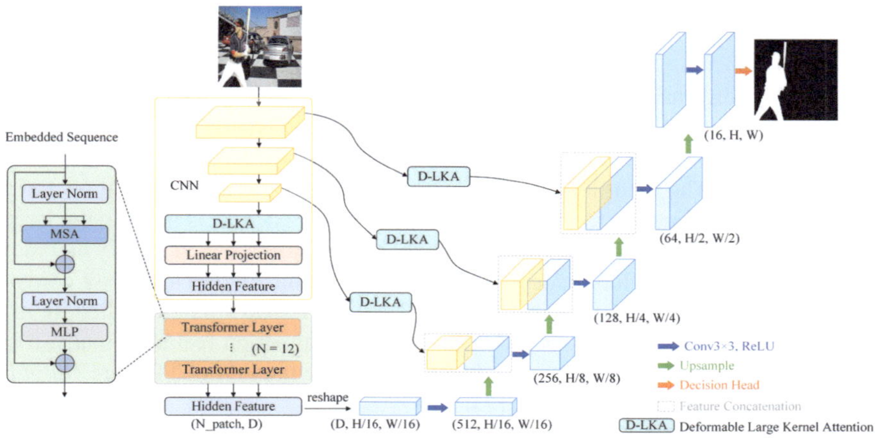

Fig. 1 Overview of the proposed network

2.1 *Hybrid Structure Encoder*

We first use CNN to extract local features from the input image to increase the overall fitting ability of the model. D-LKA is then used to refine the local features extracted by the CNN, enhancing the Transformer's ability to capture image-specific features. Finally, the feature map extracted by CNN is input into Transformer to calculate long-distance dependencies and obtain global feature representation. For the input image $x \in \mathbb{R}^{C\times H\times W}$, first through the CNN part in the hybrid structure encoder, four different scale feature maps $e_1 \in \mathbb{R}^{C_1\times\frac{H}{2}\times\frac{W}{2}}$, $e_2 \in \mathbb{R}^{C_2\times\frac{H}{4}\times\frac{W}{4}}$, $e_3 \in \mathbb{R}^{C_3\times\frac{H}{8}\times\frac{W}{8}}$, $e_4 \in \mathbb{R}^{C_4\times\frac{H}{16}\times\frac{W}{16}}$ are obtained, where C_1, C_2, C_3, C_4 are 64, 256, 512 and 1024 respectively. We selects the feature maps of the first three scales e_1, e_2, e_3 to supervise the upsampling process of the decoder and e_4 serve as the input of the subsequent Transformer part.

Before inputting to Transformer, we first input e_4 into D-LKA to get $\hat{e}_4$, which increases the spatial and channel adaptability of the input features. Then reshape e_4 into a series of 2D patches: $x_p^i \in \mathbb{R}^{P^2\cdot C}, i = 1, ..., N$. Where each patch is of size $P = 1$ and N is the number of image patches. The vectorized patches are then flattened and passed into a linear embedding layer with output dimension D. To encode the patch spatial information, we learn specific position embeddings which are added to the patch embeddings to retain positional information. The Transformer encoder consists of 12 layers of Multi-head Self-Attention (MSA) and Multi-Layer Perceptron (MLP) blocks. As shown in Fig. 1, each layer consists of an MSA module, followed by a two-layer MLP with GELU nonlinearity in the middle. A layer with layernorm (LN) is set before the MSA module and MLP, and residual connections are done after them. Finally, the output of the hybrid structure encoder is $z_L \in \mathbb{R}^{N\times D}$.

2.2 Deformable Large Kernel Attention

We combine deformable convolution [11] and large kernel attention [12], and design a deformable large kernel attention module to refine the features of different scales at each stage of the CNN part in the hybrid structure encoder to obtain more discriminative features. The input feature map F first undergoes a 1×1 convolution and a GELU activation function to obtain the feature map F'. Then F' is input to the large kernel attention part embedded with deformable convolution, which mainly includes deformable depth convolution, deformable depth hole convolution and 1×1 convolution, and the feature map Att is obtained. The feature map F is multiplied by Att and then input into a 1×1 convolution to obtain the attention feature. Finally, it is added to the original input feature map F to obtain the final refined feature $Output$. The whole process can be expressed as:

$$Att = Conv_{1\times 1}(\text{DDW - D - Conv}(\text{DDW - Conv}(F'))) \tag{1}$$

$$Output = Conv_{1\times 1}(Att \otimes F') + F \tag{2}$$

where $F \in \mathbb{R}^{C\times H\times W}$ and $F' = \text{GELU}(Conv_{1\times 1}(F))$ represent the input features, $Att \in \mathbb{R}^{C\times H\times W}$ represents the attention map and operator $\otimes$ denotes an element-wise multiplication.

2.3 Decoder

We first feed e_i into D-LKA and get $\hat{e}_i$. We reshape the final output of the Transformer part $z_L \in \mathbb{R}^{N\times D}$ into such a two-dimensional feature map form $\mathbb{R}^{D\times \frac{H}{16}\times \frac{W}{16}}$, and then use 3×3 convolution to adjust the number of channels to obtain the feature map $d_4 \in \mathbb{R}^{512\times \frac{H}{16}\times \frac{W}{16}}$. Then put d_4 to perform upsampling to increase the size of the feature maps, these feature maps are doubled in size. Then it is fused with the features received by the skip connection and input into the convolution block to get $d_3 \in \mathbb{R}^{256\times \frac{H}{8}\times \frac{W}{8}}$. This process is repeated three times in the decoder:

$$d_i = ReLU(BN(Conv_{3\times 3}(Cat(Up(d_{i+1}),\ \hat{e}_i)))), i = 3, 2, 1 \tag{3}$$

where Up represents bilinear interpolation, Cat represents feature splicing of channel dimensions, BN represents batch normalization, and is the activation function. After 3 iterations, the resulting feature map undergoes a final round of upsampling, and the final single-channel pixel-by-pixel prediction mask $\hat{y} \in \mathbb{R}^{1\times H\times W}$ is generated by the decision head.

Table 1 Quantitative comparisons by using F1, Recall and Precision as metrics. Localization performance of compared methods on different datasets

Method	Datasets								
	CASIA v2.0 +			NIST16			COLUMBIA		
	F1	Recall	Precision	F1	Recall	Precision	F1	Recall	Precision
ManTraNet	30.99	57.68	27.53	27.65	41.93	30.14	52.01	52.27	53.98
RRU-Net	84.22	92.74	81.67	77.98	76.12	85.16	77.50	76.07	81.67
C2R-Net	67.58	80.80	58.10	55.00	66.60	46.80	69.50	61.20	80.40
CAT-Net	80.12	75.05	87.54	90.19	88.28	93.77	97.44	97.55	97.84
PSCC-Net	79.78	86.73	78.15	76.28	93.70	68.41	93.53	90.92	96.90
Ours	86.72	87.21	88.51	93.68	95.74	92.28	98.80	99.36	98.25

3 Experiment

3.1 Experimental Setup

We will analyze and evaluate the proposed method on public image splicing data sets such as CASIA v2.0 [13], NIST16 [14], and COLUMBIA [15]. We use precision, recall, and F1 score to evaluate the performance of the proposed splicing forgery detection method.

3.2 Comparing with Previous Methods

In order to compare the effectiveness of the proposed method, we select five deep learning-based detection methods ManTra-Net [16] and RRU-Net [6], C2R-Net [17], CAT-Net [7] and PSCC-Net [18] are compared with the method in this article. The experimental results are shown in Table 1. As can be seen from the table, regardless of the data set used, the method proposed in this paper achieves the best performance in F1 score. Compared with state-of-the-art methods, F1 score, precision and recall are greatly improved.

3.3 Abation Study

We conducted experiments to evaluate the impact of integrating D-LKA into a hybrid structural encoder and skip connections on model detection performance. As shown in Table 2, when D-LKA is added to both the hybrid structure encoder and skip connection, the model achieves the best results, which are improved by 3.55%, 3.08%

Table 2 Ablation study of D-LKA in encoders and skip connections. We only modify some network structures

	Encoder with D-LKA	Skip with D-LKA	F1	Recall	Precision
Ours			83.17	84.13	86.07
Ours		√	85.16	86.27	87.66
Ours	√		84.56	86.66	85.18
Ours	√	√	86.72	87.21	88.51

and 2.44% respectively compared to the time without adding it. This shows that D-LKA increases the adaptability of the feature space and channels, and at the same time improves the network's ability to model splicing areas at different scales, thereby improving the performance of image splicing detection.

4 Conclusions

In this paper, we propose an image splicing detection algorithm based on a deformable large kernel attention network. The hybrid structure encoder can effectively explore local and global anomaly clues, while using D-LKA to refine features spatially and channel-wise, increasing the detection capability of small-sized spliced regions. A large number of experiments show that the method proposed in this paper has better forensic performance than existing methods.

References

1. Goljan, M., Fridrich, J.: CFA-aware features for steganalysis of color images. In: Media Watermarking, Security, and Forensics 2015, vol. 9409, pp. 279–291. SPIE (2015)
2. Chen, M., Fridrich, J., Goljan, M., Lukás, J.: Determining image origin and integrity using sensor noise. IEEE Trans. Inf. Forensics Secur. **3**(1), 74–90 (2008)
3. Korus, P., Huang, J.: Multi-scale analysis strategies in PRNU-based tampering localization. IEEE Trans. Inf. Forensics Secur. **12**(4), 809–824 (2016)
4. Hu, X., Zhang, Z., Jiang, Z., Chaudhuri, S., Yang, Z., Nevatia, R.: Span: spatial pyramid attention network for image manipulation localization. In: Computer Vision–ECCV 2020: 16th European Conference, Glasgow, UK, August 23–28, 2020, Proceedings, Part XXI 16, pp. 312–328. Springer (2020)
5. Dong, C., Chen, X., Hu, R., Cao, J., Li, X.: Mvss-net: Multi-view multi-scale supervised networks for image manipulation detection. IEEE Trans. Pattern Anal. Mach. Intell. **45**(3), 3539–3553 (2022)
6. Bi, X., Wei, Y., Xiao, B., Li, W.: RRU-Net: The ringed residual u-net for image splicing forgery detection. In: Proceedings of the IEEE/CVF Conference on Computer Vision and Pattern Recognition Workshops. pp. 0–0 (2019)
7. Kwon, M.J., Yu, I.J., Nam, S.H., Lee, H.K.: CAT-Net: Compression artifact tracing network for detection and localization of image splicing. In: Proceedings of the IEEE/CVF Winter Conference on Applications of Computer Vision, pp. 375–384 (2021)

8. Vaswani, A., Shazeer, N., Parmar, N., Uszkoreit, J., Jones, L., Gomez, A.N., Kaiser, Ł., Polosukhin, I.: Attention is all you need. Adv. Neural Inf. Process. Syst. **30** (2017)
9. He, K., Zhang, X., Ren, S., Sun, J.: Deep residual learning for image recognition. In: Proceedings of the IEEE Conference on Computer Vision and Pattern Recognition, pp. 770–778 (2016)
10. Dosovitskiy, A., Beyer, L., Kolesnikov, A., Weissenborn, D., Zhai, X., Unterthiner, T., Dehghani, M., Minderer, M., Heigold, G., Gelly, S., et al.: An image is worth 16x16 words: transformers for image recognition at scale. arXiv preprint arXiv:2010.11929 (2020)
11. Dai, J., Qi, H., Xiong, Y., Li, Y., Zhang, G., Hu, H., Wei, Y.: Deformable convolutional networks. In: Proceedings of the IEEE International Conference on Computer Vision, pp.764–773 (2017)
12. Guo, M.H., Lu, C.Z., Liu, Z.N., Cheng, M.M., Hu, S.M.: Visual attention network. Comput. Vis. Media **9**(4), 733–752 (2023)
13. Dong, J., Wang, W., Tan, T.: Casia image tampering detection evaluation database. In: 2013 IEEE China Summit and International Conference on Signal and Information Processing, pp. 422–426. IEEE (2013)
14. Guan, H., Kozak, M., Robertson, E., Lee, Y., Yates, A.N., Delgado, A., Zhou, D., Kheyrkhah, T., Smith, J., Fiscus, J.: MFC datasets: large-scale benchmark datasets for media forensic challenge evaluation. In: 2019 IEEE Winter Applications of Computer Vision Workshops (WACVW). pp. 63–72. IEEE (2019)
15. Hsu, J., Chang, S.: Columbia uncompressed image splicing detection evaluation dataset. Columbia DVMM Research Lab 6 (2006)
16. Wu, Y., AbdAlmageed, W., Natarajan, P.: Mantra-net: manipulation tracing network for detection and localization of image forgeries with anomalous features. In: Proceedings of the IEEE/CVF Conference on Computer Vision and Pattern Recognition, pp. 9543–9552 (2019)
17. Xiao, B., Wei, Y., Bi, X., Li, W., Ma, J.: Image splicing forgery detection combining coarse to refined convolutional neural network and adaptive clustering. Inf. Sci. **511**, 172–191 (2020)
18. Liu, X., Liu, Y., Chen, J., Liu, X.: PSCC-Net: progressive spatio-channel correlation network for image manipulation detection and localization. IEEE Trans. Circuits Syst. Video Technol. **32**(11), 7505–7517 (2022)

A Self-Supervised Semantic Segmentation Framework Based on Image Inpainting

Wenlong Tang, Peng Huang, and Min Zhang

Abstract With the rapid growth of remote sensing applications in various fields, the need for accurate semantic segmentation of remote sensing images has become increasingly urgent. However, the difficulty in acquiring high-quality semantically labeled data has limited the effectiveness of traditional semantic segmentation models that rely on labeled data in remote sensing. This has resulted in a need for alternative approaches to improve the accuracy of remote sensing data analysis. This paper presents a self-supervised semantic segmentation framework based on image inpainting to tackle this challenge. The proposed framework enhances the model's generalization ability by using image inpainting as a pre-task, learning a fine pixel-level representation, and applying an adversarial training technique to locate salient pixels in the image as the regions to be restored. The accuracy of both image inpainting and semantic segmentation is further improved.

Keywords Semantic segmentation · Self-supervision · Image inpainting · Remote sensing images

1 Introduction

Compared to conventional semantic segmentation techniques [1], deep learning methods [2] can automatically extract features from large-scale data without relying on domain-specific knowledge and experience to design handcrafted features. This makes deep learning a powerful tool for semantic segmentation of remotely sensed

W. Tang (✉) · P. Huang · M. Zhang
Changsha University of Science & Technology, Changsha, China
e-mail: 21208051618@stu.csust.edu.cn

P. Huang
e-mail: 21108021522@stu.csust.edu.cn

M. Zhang
e-mail: 21208051645@stu.csust.edu.cn

J. S. Park et al. (eds.), *Advanced Multimedia and Ubiquitous Engineering*, Lecture Notes in Electrical Engineering 1475, https://doi.org/10.1007/978-981-95-1565-3_10

images. One of the pioneers in this field is FCN [3], which proposes an end-to-end fully convolutional network that does not require fully-connected layers. This network can achieve precise pixel-level prediction and accurate pixel-level image classification with adequate training data. Later researchers have proposed models such as PSPNet [4], DeepLab [5], and SegFormer [6], which have significantly improved segmentation accuracy and efficiency.

Although these methods offer the advantages of robustness and high accuracy compared to traditional physical methods, they often require a large amount of high-quality labeled data, which is practically unrealistic for semantic segmentation of remote sensing images. Pseudo-labeling can effectively address this issue. However, the quality of pseudo-labels is a concern in this process. Learning from low-quality pseudo-labels can be counterproductive to the segmentation effect, which is particularly problematic for high-resolution remotely sensed images with complex pixels. Applying the chosen pre-task to remote sensing image segmentation is not an easy task due to the complexity and high resolution of surface features captured by remote sensing images and the relatively small size of specific targets. To explore accurate discriminative representations, it is necessary to mask locally salient regions instead of seeking random inpainting throughout the image.

Self-supervised semantic segmentation methods provide an effective strategy to address the above problems, as they can combine a small amount of labeled data with a large amount of unlabelled data to improve segmentation performance and reduce the burden of image annotation. We propose a framework called SIACNet that combines adversarial training and image inpainting. Firstly, SIACNet learns complex pixel-level representations of remote sensing images using image inpainting as a pre-task. Then, it predicts the salient pixels in the image as the regions that need to be restored by comparing them with the original image in an adversarial training method. This approach continuously improves the model's capability in generating adversarial to achieve better semantic segmentation.

2 Method

Figure 1 shows the overall flow of the SIACNet framework proposed in this paper, which includes two processes: the self-supervised pre-training process and the semantic segmentation process. During the self-supervised pre-training process, the Coach network takes the original image as input and outputs a semantically aware binary mask. Meanwhile, the inpainting network uses the corrupted image from Coach and attempts to restore the original image. Given the high resolution and relatively small salient regions of remote sensing images, we have implemented an adversarial training scheme to mask semantically aware regions instead of using random inpainting. This increases the difficulty of the inpainting task and allows for the learning of more discriminative feature representations. The Coach network's loss is adversarial to the loss of the image inpainting network, which enables the Coach network to gradually learn to mask salient regions for the inpainting network. During

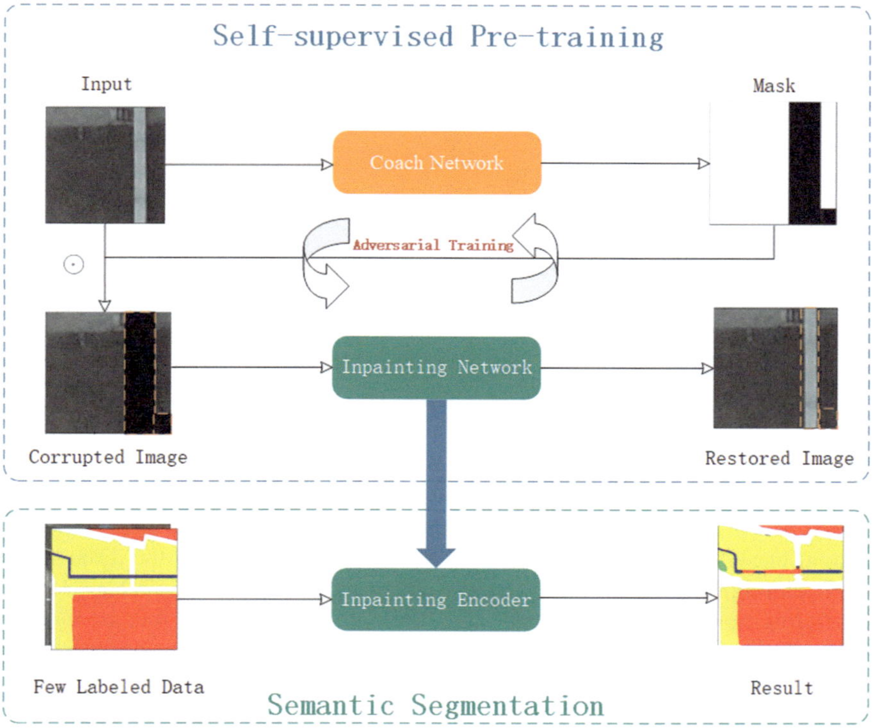

Fig. 1 The architecture of SIACNet

semantic segmentation, a small amount of labeled data is used for fine-tuning. Then, the inpainting encoder network is transferred to the semantic segmentation task, and the pixel classifiers are trained to predict the segmentation results.

2.1 *Self-Supervised Pre-Training*

In a standard image inpainting task, a random binary mask is created for each image, and pixels with a mask value of 0 are erased, while 1 is left in the image.

$$\hat{x} = M \odot x \tag{1}$$

where $\odot$ is the element-wise product operation. The inpainting network P learns to restore the image by minimizing the L_2-distance as the reconstruction loss $\mathcal{L}_{rec}$. We also add a loss function $\mathcal{L}_{con}$ for the context region to reconstruct the image more efficiently.

$$\mathcal{L}_{rec}(\hat{x}) = \frac{1}{\sum(1-M)}||(1-M)\odot(x-P(M\odot x))||_2^2 \tag{2}$$

$$\mathcal{L}_{\text{con}}(\hat{x}) = \frac{1}{\sum M}||M\odot(x-P((1-M)\odot x))||_2^2 \tag{3}$$

The definition of the total loss $\mathcal{L}_{inpainting}$ is the weighted sum of the reconstruction loss and the context loss. The weights of $\mathcal{L}_{rec}$ and $\mathcal{L}_{con}$ are represented by α_{rec} and α_{con}, respectively.

$$\mathcal{L}_{inpainting} = \alpha_{rec}\mathcal{L}_{rec} + \alpha_{con}\mathcal{L}_{con} \tag{4}$$

2.2 *Coach Network*

During a conventional image inpainting task, the Coach network does not use a random mask. Instead, it learns to mask salient regions based on the difficulty of the inpainting. The Coach network experiences the opposite of the reconstruction loss during training, resulting in the creation of corrupted images that are increasingly difficult for the inpainting network to restore. The Coach network experiences the opposite of the reconstruction loss during training, resulting in the creation of corrupted images that are increasingly difficult for the inpainting network to restore.

We propose a Coach network that learns a semantically aware corrupt mask M for image x. Coach network C identifies the most important regions in image x by maximizing the reconstruction loss in the inpainting method.

$$\mathcal{L}_{coach}(x) = 1 - \mathcal{L}_{rec}(x\odot M) = 1 - \mathcal{L}_{rec}(x\odot C(x)) \tag{5}$$

As for the selection of salient regions, the output of the Coach network is constrained to ensure that a constant proportion of the image is erased during inpainting.

2.3 *Inpainting Network*

The inpainting network employs a network structure similar to that of the Coach network. However, it does not remove the maximum pooling layer and reconstructs the image as much as possible based on the context of the image features and the masks created by the Coach network. The Coach network has a training loss that is opposite to that of the image inpainting network. Through training against this approach, the Coach network becomes more experienced at predicting significant and relevant regions as erased regions, while the inpainting network aims to restore the corrupted image as a pre-task to learn its intrinsic representation.

The Coach and inpainting networks are trained in an alternating fashion. The weights of the inpainting network remain restored while the Coach network is being trained. An adversarial training scheme is employed between the inpainting network and the Coach network, as the reconstruction loss of the former is inversely related to the loss of the latter. This means that the larger the critical masking region, the more challenging the reconstruction becomes.

During the inpainting process, the weights of the Coach network are restored. During the inpainting process, the inpainting network encoder extracts more discriminative features using the corrupted images provided by the coach network, where the critical areas are masked, increasing the difficulty of the inpainting task. The final predicted segmentation results are obtained using very few labeled images.

3 Experiment

The proposed method's effectiveness was evaluated using the ISPRS Potsdam [7] semantic labeling dataset. This dataset consists of 38 tiles, each measuring 6000 × 6000, and includes six classes: impervious surface, building, tree, low vegetation, car, and cluster/background. For pre-training, 20 tiles were used with a crop size of 600 × 600 and a stride of 200 × 200, resulting in a total of 15,680 RGB images. The testing phase was conducted on four tiles (image IDs: 2_11, 4_12, 6_7 and 7_8).

Table 1 shows the results of the comparison experiments between our proposed method and other methods on the ISPRS Potsdam dataset. The table includes three different comparison methods: (1) ImageNet [8] pre-training, where the downstream task is completed using the ImageNet pre-trained model directly; (2) training from scratch with random initialization and no pre-training; and (3) self-supervised pre-training methods. The experiments compare the proposed method with four state-of-the-art self-supervised learning methods: GLCNet [9], SimCLR [10], MultiTask [11], and MoCo v2 [12]. Each method selects mIoU and OA as evaluation metrics. Our proposed method outperforms other self-supervised methods at all percentages and outperforms methods pre-trained in ImageNet at 10% percentage and below. This proves the superiority of the proposed method when labeled data is scarce. When using only 5% of labeled data, SIACNet shows a significant improvement in accuracy over MultiTask and SimCLR, from 38.45% and 38.62% to 48.53%. However, when the percentage of labeled data is increased to 10% and above, the performance improvement of the self-supervised algorithms is not as significant as before. Despite this, SIACNet still achieves the best performance.

Table 1 Comparison of the accuracy of different methods on the Potsdam dataset

Method	Model	5%		10%	
		MIoU	OA	MIoU	OA
ImageNet	ResNet18	35.86	59.11	48.43	71.84
	ResNet50	37.43	60.47	51.23	73.48
Scratch	DeepLabv3 +	32.07	53.63	40.43	63.30
	U-Net	33.46	55.71	50.70	72.96
Self-supervised methods	MultiTask	38.45	61.22	52.56	76.65
	SimCLR	38.62	62.06	44.49	67.66
	MoCo v2	39.35	60.86	43.94	67.30
	GLCNet	39.92	62.37	44.76	66.27
	SIACNet	48.53	64.28	59.52	76.84

4 Conclusion

This paper introduces SIACNet, a new self-supervised learning technique for semantic segmentation of remote sensing images. The method uses image inpainting as a pixel-level pre-task to reduce the gap between the predecessor and downstream tasks. To enhance the discriminative representations of small salient regions and increase the difficulty of image inpainting, we propose an adversarial training scheme that comprises a Coach network and a Inpainting network. The Coach network masks the salient regions in the image, and the Inpainting network restores the corrupted image from the Coach network.

References

1. Mo, Y., Wu, Y., Yang, X., Liu, F., Liao, Y.: Review the state-of-the-art technologies of semantic segmentation based on deep learning. Neurocomputing **493**, 626–646 (2022)
2. Yuan, X., Shi, J., Gu, L.: A review of deep learning methods for semantic segmentation of remote sensing imagery. Expert Syst. Appl. **169**, 114417 (2021)
3. Long, J., Shelhamer, E., Darrell, T.: Fully convolutional networks for semantic segmentation. In: Proceedings of the IEEE Conference on Computer Vision and Pattern Recognition, pp. 3431–3440 (2015)
4. Zhao, H., Shi, J., Qi, X., Wang, X., Jia, J.: Pyramid scene parsing network. In: Proceedings of the IEEE Conference on Computer Vision and Pattern Recognition, pp. 2881–2890 (2017)
5. Chen, L.C., Papandreou, G., Kokkinos, I., Murphy, K., Yuille, A.L.: Deeplab: Semantic image segmentation with deep convolutional nets, atrous convolution, and fully connected CRFs. IEEE Trans. Pattern Anal. Mach. Intell. **40**(4), 834–848 (2017)
6. Xie, E., Wang, W., Yu, Z., Anandkumar, A., Alvarez, J.M., Luo, P.: SegFormer: simple and efficient design for semantic segmentation with transformers. Adv. Neural. Inf. Process. Syst. **34**, 12077–12090 (2021)
7. Xu, Y., Wu, L., Xie, Z., Chen, Z.: Building extraction in very high resolution remote sensing imagery using deep learning and guided filters. Remote. Sens. **10**(1), 144 (2018)

8. Krizhevsky, A., Sutskever, I., Hinton, G.E.: Imagenet classification with deep convolutional neural networks. Adv. Neural Inf. Process. Syst. **25** (2012)
9. Li, H., Li, Y., Zhang, G., Liu, R., Huang, H., Zhu, Q., Tao, C.: Global and local contrastive self-supervised learning for semantic segmentation of HR remote sensing images. IEEE Trans. Geosci. Remote Sens. **60**, 1–14 (2022)
10. Chen, T., Kornblith, S., Norouzi, M., Hinton, G.: A simple framework for contrastive learning of visual representations. In: International Conference on Machine Learning, pp. 1597–1607. PMLR (2020)
11. Li, W., Chen, H., Shi, Z.: Semantic segmentation of remote sensing images with self-supervised multitask representation learning. IEEE J. Sel. Top. Appl. Earth Obs. Remote. Sens. **14**, 6438–6450 (2021)
12. Chen, X., Fan, H., Girshick, R., He, K.: Improved baselines with momentum contrastive learning. arXiv preprint arXiv:2003.04297 (2020)

Optimizing Support Vector Machine for Dam Deformation Prediction Based on Chameleon Optimization Algorithm

Shuo Cai, Jie Zhang, and Huixin Gao

Abstract When the deformation of the dam structure reaches a critical value, a possible dam failure event may occur. Therefore, dam deformation monitoring and accurate prediction are essential. Considering that previous dam deformation prediction models have problems such as low prediction accuracy and inability to meet actual engineering needs, this paper proposes a novel method based on support vector machines to solve this problem. Specifically, the Chameleon Optimization (CO) algorithm is proposed. The CO algorithm is used to optimize the penalty factor and kernel function parameters of the Support Vector Machine (SVM), and a dam deformation prediction model based on the CO-SVM algorithm is established. Particle Swarm Optimization (PSO) algorithm and Gray Wolf Optimization (GWO) algorithm were introduced to optimize SVM to construct the PSO-SVM model and GWO-SVM model respectively. Through the measured data of Shuibuya concrete-faced rockfill dam, the prediction results are compared with the prediction results of SVM, PSO-SVM and GWO-SVM models. The final experimental results verified the effectiveness and superiority of the method.

Keywords Structural health monitoring · Settlement prediction · Optimization algorithm · SVM

1 Introduction

Dam failure will have a serious impact on people's lives and property. To timely judge the operation of the dam, a large number of monitoring instruments are embedded in the dam to monitor environmental variables and structural response. Establishing a suitable deformation prediction model based on measured data can help ensure the long-term safe operation of the dam [1].

S. Cai (✉) · J. Zhang · H. Gao
School of Computer and Communication Engineering, Changsha University of Science and Technology, Changsha, China
e-mail: caishuo@csust.edu.cn

J. S. Park et al. (eds.), *Advanced Multimedia and Ubiquitous Engineering*, Lecture Notes in Electrical Engineering 1475, https://doi.org/10.1007/978-981-95-1565-3_11

Regression statistical models are often used in dam deformation prediction modeling, but they cannot effectively solve the multicollinearity problem existing in input variables and dam deformation [2]. Support Vector Machine (SVM) overcomes these shortcomings and has obvious advantages in solving nonlinear problems with small samples and high dimensions [3]. However, the prediction accuracy of SVM depends on the appropriate settings of hyperparameters. This study proposes the Chameleon Optimization (CO) algorithm and uses it to search for the best parameters of the SVM model, and constructs a dam deformation prediction model based on the CO algorithm-optimized SVM. The model has achieved good results in engineering applications.

2 Method

The Chameleon Optimization (CO) algorithm introduced in this article is inspired by the adaptability and color changes of chameleons. The algorithm simulates the behavior of a chameleon during social interactions, camouflaging itself by changing colors in social situations to gain favor from the opposite sex. The algorithm can be described as:

Let $f(x)$ be the objective function to be optimized. In the initial stage, some solutions $(x_1, x_2 \cdots x_n)$ are randomly generated in the search space, each solution represents an individual, and n is the population size. The merits of these solutions are evaluated by calculating individual fitness values. The individual constantly updates its position to approach the optimal individual position. The mobile strategy adopted by the CO algorithm in the social process can be expressed as:

$$x_{new} = x + rM \tag{1}$$

where x_{new} is the new position after movement; x represents the current position; M is the step size; and r is a random number sampled from a uniform distribution (-1, 1). The expression of the step length can be described as:

$$M = \left(bound_{up} - bound_{lp}\right) * \beta \tag{2}$$

where $bound_{up}$ and $bound_{lp}$ are the upper and lower bounds of the search space, and β is the step adjustment factor.

In the CO algorithm, the colors between individuals can reflect the degree of difference between them. Individuals with lower fitness adjust their color slightly, while individuals with higher fitness adjust their color substantially. Use the following formula to make color adjustments:

$$C = \alpha(x_{best} - x) \tag{3}$$

where C is the expression of the color depth of the chameleon individual, x_{best} is the position of the optimal individual, and α is the color adjustment factor. Fine-tune position based on individual color, expressed by the following formula:

$$x = x + aC_{new} \tag{4}$$

where a is a random number in (0, 1).

Each movement requires boundary constraints on the updated individual positions to ensure that they are within the allowed range. The fitness value of the individual is then recalculated based on the new location and compared with the previous fitness value. If the new fitness value is better, the individual's fitness value is updated, otherwise, it remains unchanged. Finally, through continuous iteration of position movement and color adjustment, the CO algorithm can search for a better solution in the solution space.

3 CO-SVM Model Construction

The penalty factor (C) and kernel parameter (γ) of the SVM model have a great impact on its prediction performance [4]. This paper proposes a new hybrid SVM model that uses the CO algorithm to select optimal SVM parameters and uses the mean square error (MSE) of the model as the objective function. The specific implementation steps of dam deformation prediction based on CO-SVM are as follows shown in Fig. 1.

(1) Receive dam monitoring data, select effective influencing factors (feature variables) as input, and dam deformation as output;
(2) Preprocess the selected data, including outlier removal, data interpolation and data normalization;
(3) Divide the data set, divide the first 80% of the preprocessed data into a training set, and the last 20% into a test set;
(4) Build the SVM model and select the parameters C and γ to be optimized;
(5) Initializing parameters in the optimization algorithm and C and γ of SVM;
(6) Initialize the population position and randomly generate individuals in the search space according to the constraints of the chameleon population size and spatial dimensions;
(7) Apply Eq. (8) to update individual positions, record the optimal position as x_{best};
(8) Calculate the objective function to obtain the fitness values of all individuals;
(9) Update the optimal individual position;
(10) Judge the individual fitness value, apply Eq. (10) to update the color of the chameleon for individuals with high fitness values, and apply Eq. (11) to fine-tune the individual position according to the depth of the color;

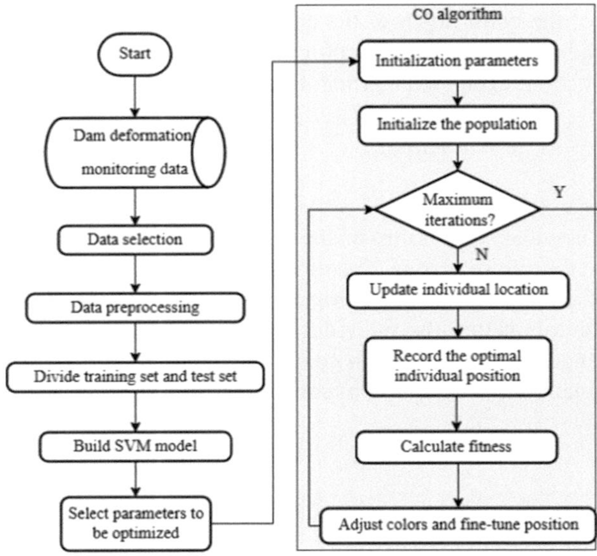

Fig. 1 Construction process of dam deformation prediction model based on CO-SVM

(11) Reach the maximum number of iterations, the global optimal solution is output as the optimal parameters;
(12) Input the optimal parameter combination into the SVM prediction model to construct a dam deformation prediction model based on CO algorithm-optimized SVM;
(13) Compare and analyze predicted values with actual values, and use statistical indicators to evaluate the performance of the established model.

The predictive performance and accuracy of the model are measured by several statistical indexes, including mean absolute error (MAE), root mean square error (RMSE), and correlation coefficient (R). The best model should have the smallest MAE and RMSE values and R close to 1.

4 Case Study

4.1 Description of the Gravity Dam

In this study, the Shuibuya concrete-faced rockfill dam was selected as an engineering case. The dam is located in Hubei Province, China. The effectiveness of the CO-SVM model is verified using SA1-9 measured data. The vertical displacement of the dam is measured once a month on average. This study selects 147 consecutive sets of

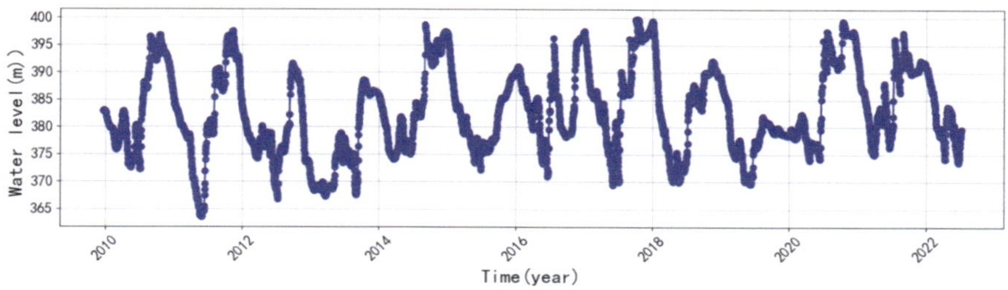

Fig. 2 Upstream water level

recorded data from December 20, 2009, to July 7, 2022, as the research object. The cumulative vertical displacement is shown in Fig. 3. During this period, the water level upstream of the dam is shown in Fig. 4. The first 80% is used as the training set for the experiment, and the last 20% is used as the test set (Fig. 2).

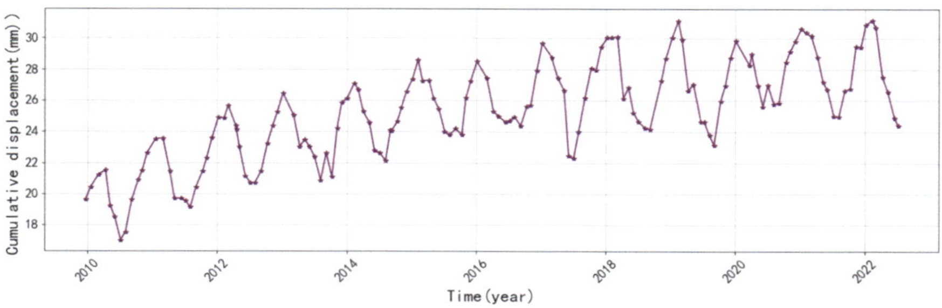

Fig. 3 Accumulated settlement at the measurement point

Fig. 4 Relation curves between fitness value and iteration number

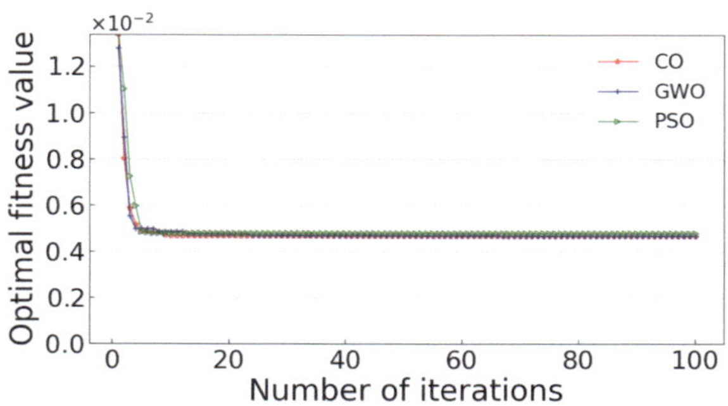

4.2 Effective Factors

The HST model [5] is currently the most popular data-based dam monitoring and analysis method. The HST model divides the influencing factors of dam deformation into water pressure components, temperature components, and aging components. The input vector of the dam deformation prediction model is

$$x = \left\{H^1, H^2, H^3, \sin(s), \cos(s), \sin(2s), \cos(2s), \theta, \ln(\theta + 1)\right\} \tag{5}$$

where H is the upstream water level; $s = 2\pi t/365$, $\theta = t/100$, and t is the cumulative number of days since the date of analysis. Since there is a lot of noise in the original monitoring data of the dam, the data is preprocessed before being input into the model, including the three steps of outlier removal, interpolation, and normalization.

4.3 Model Parameter Settings

Before running the model, the initial parameters are set as follows. The maximum number of iterations is set to 100, the population size is 10, the range of the penalty factor C and the kernel parameter γ is set to [0.01, 10]. Particle swarm optimization algorithm and gray wolf optimization algorithm were introduced to construct PSO-SVM and GWO-SVM dam deformation prediction models as comparison models. In the chameleon optimization algorithm, set the color factor to 0.5 and the step adjustment factor to 0.1. In the particle swarm optimization algorithm, the acceleration factors $c1$ and c2 are set to 1, and the inertia weight is set to 0.8. The relationship between fitness and the number of iterations of the three algorithms is shown in Fig. 4. As can be seen from Fig. 4, the CO algorithm can also find the optimal solution quickly and has good convergence accuracy.

4.4 Dam Deformation Prediction Results and Analysis

The statistical indicators of the prediction results of SVM, PSO-SVM, GWO-SVM and CO-SVM models are shown in Table 1. The comparison between the prediction results of each model and the actual values is shown in Fig. 5a, b shows their residuals. As can be seen from Table 1, the CO-SVM model has the smallest MAE and RMSE, and largest R. It can be seen from Figs. 5 and 6 that the CO-SVM model has strong fitting ability. The results show the applicability of the CO algorithm-optimized SVM model in dam deformation prediction., and the CO algorithm proposed in this article has certain advantages in parameter optimization.

Table 1 Performance of each model

Prediction model	MAE	RMSE	R
SVM	0.5725	0.7292	0.9327
PSO-SVM	0.3728	0.4840	0.9710
GWO-SVM	0.3769	0.4861	0.9707
CO-SVM	0.3695	0.4832	0.9737

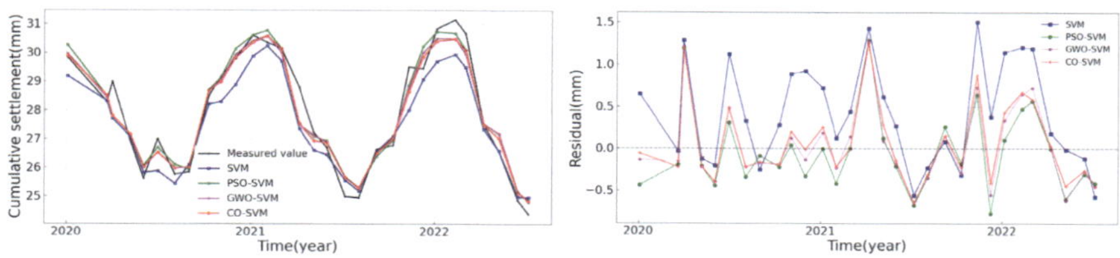

(a) (b)

Fig. 5 Prediction results and residuals of each model: **a** prediction results; **b** residuals

5 Conclusions

This paper proposes a chameleon optimization algorithm based on motion strategy and color change, uses this algorithm to optimize SVM, and constructs a dam deformation prediction method based on the CO-SVM model. Taking the monitoring data of Shuibuya concrete-faced rockfill dam as an example, the prediction results of SVM, PSO-SVM, GWO-SVM and CO-SVM models were compared and analyzed. The conclusion drawn from this is that the CO algorithm proposed in this article has certain advantages in optimizing SVM model parameters, and the CO-SVM dam deformation prediction model constructed has high prediction accuracy.

References

1. D Yuan C Gu X Qin 2022 Performance-improved TSVR-based DHM model of super high arch dams using measured air temperature Eng. Struct. 250 113400
2. C Shao C Gu M Yang 2018 A novel model of dam displacement based on panel data Struct. Control. Health Monit. 25 1 e2037
3. H Su X Li B Yang 2018 Wavelet support vector machine-based prediction model of dam deformation Mech. Syst. Signal Process. 110 412 427
4. Jun, Z.: The development and application of support vector machine. J. Phys. Conf. Ser. **1748**(5), 052006 (IOP Publishing) (2021)
5. X Yu J Li F Kang 2023 A hybrid model of bald eagle search and relevance vector machine for dam safety monitoring using long-term temperature Adv. Eng. Inform. 55 101863

Optimization of MaxSAT Local Search Solver Based on J-init Assignment

Chao Xu and Kang Liu

Abstract Maximum satisfiability (MaxSAT) is a well-known optimization problem and has a wider range of practical applications. Local search is an efficient incomplete method for solving MaxSAT and is useful in fields where high-quality solutions need to be obtained in a reasonable time. For the local search MaxSAT solver, the effectiveness of the initial assignment affects the actual performance, and the existing random initial assignments are often quite different from the optimal solution. In this work, we propose a structure-based initial assignment method, which will obtain an effective initial assignment according to the specific structure of the MaxSAT, called J-init. Experimental results show that the initialization method based on the J-init can effectively improve the solution quality of the MaxSAT problem within the specified time. By using this method, we can get a reasonable initial assignment in a shorter time and provide a more promising starting point for the subsequent optimization process.

Keywords MaxSAT · Local search · Johnson algorithm

1 Introduction

The Maximum Satisfiability problem (MaxSAT) is an optimized version of the Satisfiability problem (SAT). The Satisfiability Problem of propositional logic is one of the most important problems in computer science, and it is also the first problem that has been proven to be NP-complete (NPC) in the world [1].

Both MaxSAT and SAT are based on the Conjunctive Normal Form (CNF) of propositional logic and constraint models formed by setting constraints and solving goals. The goal of MaxSAT is to find a solution such that a given CNF satisfies

C. Xu · K. Liu (✉)
Changsha University of Science and Technology, Changsha, China
e-mail: laka@stu.csust.edu.cn

C. Xu
e-mail: laka@stu.csust.edu.cn

J. S. Park et al. (eds.), *Advanced Multimedia and Ubiquitous Engineering*, Lecture Notes in Electrical Engineering 1475, https://doi.org/10.1007/978-981-95-1565-3_12

the maximum number of clauses under constraints. It is a constrained optimization model.

According to the different properties of CNF solved, MaxSAT problems are divided into different categories, including Partial MaxSAT and Weighted MaxSAT. Partial MaxSAT means that CNF clauses are divided into hard clauses and soft clauses. Its goal is to find an assignment that satisfies all hard clauses and maximizes the number of satisfied soft clauses. Weighted MaxSAT, on the other hand, adds a positive integer multiple of the Weight to each clause in CNF, and the goal is to find a satisfying clause with the largest weight and assignment. Many real-world problems can be modeled more accurately by MaxSAT instead of SAT, such as the timetables problem [2] and the maximum clique problem [3]. Therefore, it is of great theoretical and practical significance to study effective methods for solving MaxSAT problems.

Local search is an efficient algorithm for solving MaxSAT. The local search solver to maintain a complete the assignment, and iteration to modify the assignment (i.e., flip the value of the variable). They keep track of the best solution found throughout the search and return it when a termination criterion, such as a time limit, is reached [4]. Recent work has shown that the mechanism for generating initial assignments substantially affects the performance of local search [5].

The Johnson algorithm [6] iteratively constructs assignments by traversing the variables in a heuristic variable order that we specify. In previous work, Berend et al. attempted to solve MaxSAT problems by combining the Johnson algorithm with configuration checking local search (CCLS) [7]. Experimental results showed that the Johnson algorithm enabled CCLS to find high-quality feasible or optimal solutions with fewer flips. However, the Johnson algorithm did not fully consider the structure of the problem instances. Therefore, we propose an improvement called the J-init algorithm. J-init obtains initial solutions that closely match the structure of MaxSAT instances and demonstrates improved performance when combined with several advanced local search solvers.

2 Our Main Contributions

Inspired by the application of the Johnson algorithm in various domains. In this work, we propose an initial assignment method to optimize MaxSAT solving, called J-init. J-init scores clauses and literals based on the structure of the instance and probabilistically assigns values to variables using the positive and negative scores. This assignment is passed to a local search solver as the initial allocation for solving. In essence, J-init is a generic approach, as it can be combined with any local search solver.

To demonstrate the generality of J-init, we conducted extensive experiments on all incomplete track instances from the MaxSAT Evaluations (MSEs) over the past three years. We integrated and compared J-init with the three most prominent local search solvers: SATLike3.0 [4], BandMaxSAT [8], and NuWLS [9]. Our research findings indicate that J-init significantly enhances the quality of solutions across various local search solvers and reduces the time required to find feasible solutions. This underscores J-init's potential to advance the development of local search solvers to a certain extent.

3 Preliminary

Given a set of n Boolean variables $\{x_1, x_2, ..., x_n\}$, a literal is either a variable x_i or its negation $\neg x_i$. A clause is a disjunction of literals, $i.e.$, $c_i = l_{i,1} \vee l_{i,2} \vee ... \vee l_{i,k_i}$, where $l_{i,j}$ is a literal and k_i denotes the length of clause c_i. .. A formula in conjunctive normal form (CNF), $i.e.$, $F = c_1 \wedge c_2 \wedge ... \wedge c_m$, is a conjunction of clauses.

An assignment is a mapping that assigns a Boolean value (0 or 1) to each variable. Given an assignment T, each clause is either satisfied or unsatisfied: a clause c_i is satisfied if at least one of the literals in c_i is true under T; herwise, c_i is unsatisfied. An assignment T is feasible if it satisfies all hard clauses; Otherwise, the assignment T is not feasible. Usually, the quality of a solution T is judged by $cost(T)$. For partial MaxSAT, $cost(T)$ represents the sum of the weights of soft clauses not satisfied by the solution T, and for MaxSAT, $cost(T)$ represents the number of soft clauses not satisfied by the solution T.

4 Method of Initial Assignment

4.1 Local Search Solver

Local search solvers typically start with a random initial assignment and iteratively flip literals in unsatisfied clauses. They then evaluate whether the flipped assignment is feasible. If it is, they compare s cost with the previous feasible solution, favoring the one with a smaller cost as the current best solution. When the specified time limit is reached, the solver returns the best solution found. Notable advanced local search solvers include SATLike3.0, BandMaxSAT, and NuWLS.

We use J-init to initialize the initial assignment of the local search. To prevent getting stuck in local optima, we assign the value obtained from J-init with a 90% probability; otherwise, we assign the opposite value.

4.2 Algorithm of J-init

The j-init algorithm determines the probability of each variable taking 0 or 1 by scoring each variable. In a CNF formula with n variables and m clauses. The weight of the literal l_i is determined by the formula: $\sum_{i=1}^{r} w_j/2^{|C_j|}$.where $1 \leq i \leq n$ and $1 \leq j \leq m$. r is the number of clauses in which the literal is located. w_j is the weight of the clause j. It is worth indicating that the weighting of each soft clause in the weighted MaxSAT formula. For hard clauses, we set the weight of the hard clause to the sum of the scores of all soft clauses, so that the basic requirement that the hard clause is satisfied first is also satisfied. Unweight MaxSAT, on the other hand, fixes w_j to 1.

The improvements to J-init over Johnson algorithm can be summarized as follows (Algorithm 1 demonstrates this process, the time complexity is $O(l)$):

- When comparing the scores of positive and negative literals, we maintain a probability p, where $p = w_{l^+}/(w_{l^+} + w_{l^-})$. Here, w_{l^+} represents the score of positive literals, w_{l^-} represents the score of negative literals, and p represents the probability that the variable takes the value 1. We use this probability to assign values to variables.
- After obtaining scores for all variables, we directly assign variables with $p > 0.8$ to TRUE and those with $p < 0.2$ to FALSE. We propagate the assigned variables.
- For variables with equal scores for positive and negative literals, we compute the sum of scores for the neighbors of both positive and negative literals and use this score for probability-based assignment.
- Variables with identical probabilities p are assigned the same value because they exhibit similar or nearly identical structural characteristics.

Algorithm 1: initial solution algorithm with J-init

INPUT: A CNF formula F
OUTPUT: A truth assignment T

```
T, p ← φ;
for each variable x_i do
    w_{l_i^+}, w_{l_i^-} ← Σ_{i=1}^{r} w_j/2^{|C_j|};
    p_i ← w_{l_i^+} / (w_{l_i^+} + w_{l_i^-})
```

(continued)

(continued)

Algorithm 1: initial solution algorithm with J-init

    if $p_i > 0.8$ **then**
        $T_i \leftarrow 1$;
        propagate $x_i = 1$; //Unit Propagation
    end if
    If $p_i < 0.2$ **then**
        $T_i \leftarrow 0$;
        propagate $x_i = 0$;
    end if
end for
for each value p_i **do**
    if there exists j such that $|p_i - p_j| < 0.1$ **then**
        $T_i \leftarrow T_j$;
    end if
    if $p_i == 0.5$ **then**
        $w_{n_i^+}, w_{n_i^-} \leftarrow \sum_{i=1}^{r} w_{l_i^n}$; //$w_{l_i^n}$ is the neighbour of l
        $p_i \leftarrow w_{n_i^+} / \left(w_{n_i^+} + w_{n_i^-} \right)$;
    end if
    if T_i is *NULL* **then**
        Take p_i as the probability to take 1 or 0 for T_i
    end if
end for
return T

5 Experimental Study

5.1 Testing Methodology

Our experiments were conducted on an Ubuntu system with 256G of RAM with Hygon CPUs 7185@2.0 GHz. All experiments are evaluated on incomplete track instances from MaxSAT Eval-uations 2021–2023. In our experiments, tested on three state-oftheart local search solvers, namely SATLike3.0, BandMaxSAT, NuWLS. Each solver is executed once on each instance, with a time limit of 300 s and 60 s. We use SL + , BM + , and NL + to represent SATLike, BandMaxSAT, and NuWLS using the J-init algo-rithm. The solver based on the original Johnson algorithm is abbreviated as SL− ,BM −, NL −. The original solver was called SL, BM, NL.

5.2 Experimental Results

Table 1 shows the comparison in the number of solutions between the J-int solver and the original solver within 300 s. "W" indicates a weighted basis and "UW" indicates an unweighted basis. #inst represents the number of benchmarks. It can be seen from the table that in most cases, the solver using J-init can solve more instances in a certain period of time than the original solver.

Table 2 shows the time required to find the final feasible solution based on the J-init solver and the original solver within 300 s. It is not difficult to see that the J-init method shortens the distance between the initial solution and the final feasible solution. Reduces the time to find feasible solutions.

The results in Table 3 show that the J-init algorithm based on the formula structure is effective in improving Johnson, improving the performance of the Johnson algorithm and being able to find feasible solutions for more instances in a short time.

Table 1 Comparison of the number of solutions within 300 s

Benchmark	#inst	SL	SL +	BM	BM +	NL	NL +
W_2021	141	134	130	132	**133**	121	**132**
W_2022	197	179	**180**	174	171	168	**182**
W_2023	160	129	**143**	149	142	132	**144**
UW_2021	136	120	120	117	**121**	122	**123**
UW_2022	179	171	164	153	**169**	149	**152**
UW_2023	179	163	159	162	157	142	**149**

Table 2 Comparison of average time within 300 s

Benchmark	#inst	SL	SL +	BM	BM +	NL	NL +
W_2021	141	109.34	**99.23**	112.78	**105.45**	116.89	118.56
W_2022	197	131.22	133.58	135.78	**134.24**	125.63	**119.76**
W_2023	160	134.59	**134.16**	137.41	**136.27**	130.89	**123.03**
UW_2021	136	119.59	**117.13**	115.44	**111.33**	118.71	**101.36**
UW_2022	179	131.73	**128.70**	135.61	136.70	123.49	124.17
UW_2023	179	128.98	**128.64**	131.90	**128.44**	114.23	**113.76**

Table 3 Comparison between the number of solutions and Johnson algorithm within 60 s

Benchmark	#inst	SL −	SL +	BM −	BM +	NL −	NL +
W_2021	141	124	**126**	92	**126**	128	**129**
W_2022	197	155	**165**	154	**169**	174	174
W_2023	160	140	**152**	148	**149**	151	**155**
UW_2021	136	126	126	126	126	126	126
UW_2022	179	166	**168**	165	**167**	167	**168**
UW_2023	179	159	**164**	155	**161**	160	**167**

6 Conclusion

In this work, we propose an initial assignment method based on the J-init algorithm for use as the initial solution in MaxSAT local search solvers. Experiments show that our approach significantly improves the performance of three state-of-the-art local search solvers, confirming its generality. For future research, we plan to integrate this method with complete solvers, aiming to enhance the time to obtain optimal solutions.

Acknowledgements This paper was supported by Supported by the National Natural Science Foundation of China under Grants (62002032).

References

1. Cook S.A.: The complexity of theorem-proving procedures. In: Proceedings of the Third Annual ACM Symposium on Theory of Computing, pp. 151–158 (1971)
2. Emir, D., Nysret, M.: Maxsat-based large neighborhood search for high school timetabling. Comput. Oper. Res. **78**, 172–180 (2017)
3. Fang, Z.W., Li, C.M., Xu, K.: An exact algorithm based on maxsat reasoning for the maximum weight clique problem. J. Artif. Intell. Res. **55**, 799–833 (2016)
4. Cai, S.W., Lei, Z.D.: Old techniques in new ways: clause weighting, unit propagation and hybridization for maximum satisfiability. Artif. Intell. **287**, 103–354 (2020)
5. Zhang, W.J., Sun, Z.Y., Zhu, Q.H., Li, G., Cai, S.W., Xiong, Y.F., Zhang, L.: NLocalSAT: boosting local search with solution prediction. Proceedings of IJCAI **2020**, 1177–1183 (2020)
6. Johnson D.S.: Approximation algorithms for combinatorial problems. In: Proceedings of the Fifth Annual ACM Symposium on Theory of Computing, pp. 38–49 (1973)
7. Berend, D., Golan, S., Twitto, Y.: Using the method of conditional expectations to supply an improved starting point for CCLS. J. Comb. Optim. **44**(5), 3711–3734 (2022)
8. Zheng, J.Z., He, K., Zhou, J.R., Jin, Y., Li, C.M., Felip, M.: Bandmaxsat: a local search maxsat solver with multi-armed bandit. In: The Thirty-First International Joint Conference on Artificial Intelligence, pp. 1901–1907 (2022)
9. Chu, Y., Cai, S., Luo, C.: NuWLS: improving local search for (weighted) partial MaxSAT by new weighting techniques. Proc. AAAI Conf. Artif. Intell. **37**(4), 3915–3923 (2023)

A Semi-supervised Semantic Segmentation Framework Based on Consistency Regularization

Wenlong Tang, Min Zhang, and Peng Huang

Abstract With the rapid growth of remote sensing applications in various fields, the need for accurate semantic segmentation of remote sensing images has become increasingly urgent. However, the difficulty in acquiring high-quality semantically labeled data has limited the effectiveness of traditional semantic segmentation models that rely on labeled data in remote sensing. This has resulted in a need for alternative approaches to improve the accuracy of remote sensing data analysis. We presents a semi-supervised semantic segmentation framework based on consistency learning. On the one hand, a unified perturbation strategy that unifies the image-level perturbation and the feature-level perturbation into separate streams to exploit a larger perturbation space. On the other hand, a dual-stream perturbation strategy is designed to fully explore the predefined image-level perturbation space and take advantage of comparative learning for discriminative characterization. The framework integrates the above two strategies and shows superior performance for semantic segmentation tasks in the case of remote sensing images with sparsely labeled data.

Keywords Semantic segmentation · Semi-supervision · Consistency regularization · Remote sensing images

1 Introduction

The latest techniques for semi-supervised semantic segmentation have developed from adversarial training methods based on GANs [1] to widely adopted consistent regularization frameworks and emerging self-training methods. These techniques

W. Tang (✉) · M. Zhang · P. Huang
Changsha University of Science and Technology, Changsha, China
e-mail: 21208051618@stu.csust.edu.cn

M. Zhang
e-mail: 21208051645@stu.csust.edu.cn

P. Huang
e-mail: 21108021522@stu.csust.edu.cn

J. S. Park et al. (eds.), *Advanced Multimedia and Ubiquitous Engineering*, Lecture Notes in Electrical Engineering 1475, https://doi.org/10.1007/978-981-95-1565-3_13

follow the research line of Semi-Supervised Learning (SSL) [2]. Our work will focus on consistent regularization frameworks [3] that range from weak to strong. These frameworks were generalized by FixMatch [4] in the field of semi-supervised classification and subsequently influenced many other related tasks.

FixMatch degenerates into an inexperienced online self-training method that produces worse results without image-level perturbations. However, these perturbations are limited to the image level, which restricts the model from exploring a wider range of perturbations and maintaining consistency across levels.

FixMatch is a powerful framework for semantic segmentation scenarios. A reproduced FixMatch outperforms almost all existing methods for this task. We propose a unified perturbation framework based on FixMatch, which separates image-level perturbation and feature-level perturbation into separate streams to exploit a wider perturbation space. On the other hand, a dual-stream perturbation strategy is designed to fully explore the predefined image-level perturbation space and utilize comparative learning for discriminative characterization. Our proposed framework, CDPMatch, integrates the above two components and outperforms existing methods in both segmentation metrics for the Potsdam dataset. Notably, it also demonstrates strong advantages in medical image analysis and remote sensing interpretation.

2 Method

As mentioned above, FixMatch exploits unlabeled data using weak-to-strong consistent regularization. Specifically, each unlabeled image x is simultaneously perturbed by two operators, i.e. weak perturbation $\mathcal{A}^w$ (e.g. cropping) and strong perturbation $\mathcal{A}^s$ (e.g. color dithering). Then the total objective function is the combination of the supervised loss $\mathcal{L}_s$ and the unsupervised loss $\mathcal{L}_u$, i.e.:

$$\mathcal{L} = \frac{1}{2}(\mathcal{L}_u + \mathcal{L}_s) \tag{1}$$

2.1 IFPerb

Image-level perturbations, such as color dithering and CutMix, introduce heuristic biases that provide additional a priori information to FixMatch's bootstrapping paradigm. This captures the benefits of consistent regularization. However, it is important to note that these approaches are limited to the image level, which restricts the model's ability to handle multi-level consistency for more diverse perturbations. To construct a broader perturbation space based on FixMatch, we propose injecting feature perturbations for a weakly perturbed image x^w. We suggest dividing different levels of perturbations into multiple independent feedforward streams, enabling the student model to achieve target consistency more directly in each stream. In FixMatch, we not only obtain p^w and p^s but also p^{fp} from the auxiliary feature perturbation streams by:

$$e^w = g(x^w) \tag{2}$$

$$p^{fp} = h(\mathcal{P}(e^w)) \tag{3}$$

where e^w is the extracted feature of x^w and $\mathcal{P}$ denotes feature perturbation such as filtering or adding uniform noise.

As shown in Fig. 1a, each unlabeled small batch of data contains three feedforward streams: (1) the simplest stream: $x^w \to f \to p^w$; ; (2) the image-level strong perturbation stream: $x^s \to f \to p^s$; ; and (3) the feature perturbation stream: $x^w \to g \to \mathcal{P} \to h \to p^{fp}$, which we call IFPerb. This ensures that the student model is consistent with other perturbations at the image and feature level.

$$\mathcal{L}_u = \frac{1}{B_u} \sum 1(\max(p^w) \geq \tau)(H(p^w, p^s) + H(p^w, p^{fp})) \tag{4}$$

It should be noted that the module does not aim to present a new approach to feature perturbation in this work. Instead, the module employs a simple channel dropout (nn.Dropout2d in PyTorch). Dropout does not regularize activations, in which case nn.Dropout2d( ) helps to promote independence between feature mappings and also has the effect of preventing model overfitting.

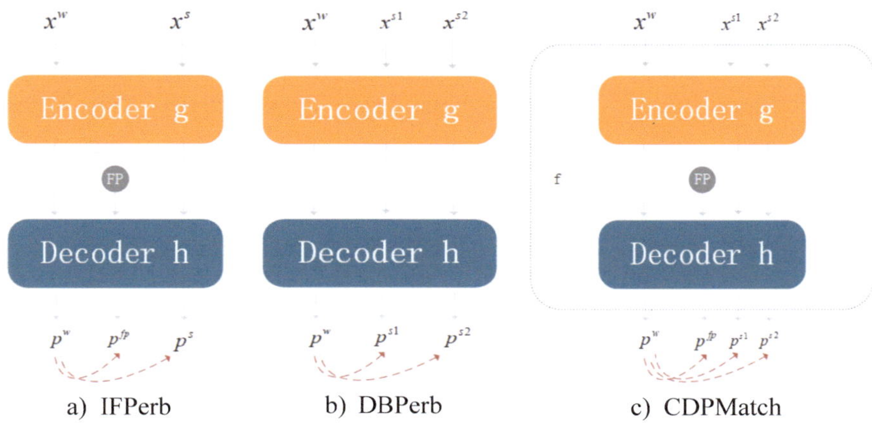

Fig. 1 The structure of IFPerb, DBPerb and CDPMatch

2.2 DBPerb

Motivated by the significant benefits of image-level strong perturbations, we aim to thoroughly investigate them based on the previous work. Our inspiration comes from recent advancements in the fields of self-supervised learning and semi-supervised classification, which suggest that constructing multiple views with unlabeled data as input can better exploit perturbations. For instance, SwAV [5] proposes a new technique called multi-crop that can achieve local-to-global consistency between multiple views with varying resolutions. Similarly, ReMixMatch [6] generates several robust data-enhanced versions for model learning.

Therefore, we investigated whether a simple idea could improve semi-supervised semantic segmentation. Our proposed module generates a two-stream perturbation (x^{s1}, x^{s2}) from x^w using a strong perturbation pool $\mathcal{A}^s$, instead of feeding a single p^s into the model. As $\mathcal{A}^s$ is predefined but non-deterministic, x^{s1} and x^{s2} are not equal. This idea originates from contrastive learning in self-supervised learning methods. It involves generating two similar views through different data augmentations, which are sampled from the same data augmentation series. As a result, the two augmented views are extremely similar. However, our proposed two-stream perturbation frameworks generate views obtained through strong perturbations (e.g. color dithering) which is used in the dual stream perturbation framework (DBPerb) shown in Fig. 1b. The perturbations are stochastically non-deterministic, resulting in different generated views. The framework aims to reduce bias and maintain objectivity.

Inspired by related work, we decided to regularize the two strong views with a shared weak view graph. This enforces consistency between the two strong views across platforms. In the cross-entropy loss employed by the module, k^w is the classifier weight of the category predicted by x^w and (q_{s1}, q_{s2}) is a feature of image (x^{s1}, x^{s2}). We maximize $q_j \cdot k_w$ for $\sum_{i=0}^{C} q_j \cdot k_i$, where $j \in \{s_1, s_2\}$ and k_i are the classifier weights of category i. Therefore, this module satisfies the InfoNCE loss [7]:

$$\mathcal{L}_u = \frac{1}{B_u} \sum \mathbb{1}(\max(p^w) \geq \tau) \cdot (\lambda \mathrm{H}(p^w, p^{fp}) + \frac{\mu}{2}(\mathrm{H}(p^w, p^{s_1}) + \mathrm{H}(p^w, p^{s_2}))) \tag{5}$$

q_{s1} and q_{s2} are positive sample pairs, while all other classifier weights except k^w are negative samples. This approach shares similarities with contrastive learning, which is useful for learning discriminative representations and is relevant for semantic segmentation tasks.

2.3 CDPMatch

In summary, this paper presents two modules for exploiting unlabeled remote sensing images: IFPerb and DBPerb. Figure 1c illustrates the overall framework for integrating these two approaches, called CDPMatch. Unlike FixMatch, our integrated framework retains two auxiliary feedforward streams, one for perturbing the features of x^w and the other for multi-view learning of (x^{s1}, x^{s2}). The two auxiliary feedforward streams are used for learning the features, respectively. The final unsupervised loss formulation is presented:

$$\mathcal{L}_u = \frac{1}{B_u}\sum \mathbb{1}(\max(p^w) \geq \tau) \cdot (\lambda \mathrm{H}(p^w, p^{fp}) + \frac{\mu}{2}(\mathrm{H}(p^w, p^{s_1}) + \mathrm{H}(p^w, p^{s_2}))) \tag{6}$$

As the feature-level perturbation flow and the image-level perturbation flow each have their distinct characteristics and advantages, their loss weights λ and μ have been equally set to 0.5. The conventional cross-entropy loss (H in $\mathcal{L}_u$) has been employed, and a confidence threshold of 0.95 (τ) has been set for all datasets.

3 Experiment

The proposed method's effectiveness was evaluated using the ISPRS Potsdam semantic labeling dataset. This dataset consists of 38 tiles, each measuring 6000 × 6000, and includes six classes: impervious surface, building, tree, low vegetation, car, and cluster/background. The experiment involved dividing the original training dataset into two subsets: a labeled data subset and an unlabeled data subset. The subsets were created by randomly selecting 1/16, 1/8, 1/4, and 1/2 of the data from the entire training dataset to form the labeled data subset, while the remaining data formed the unlabeled data subset. For model training, the Potsdam dataset was split into 24 images for training and 14 for testing. From the training set, 7 images were randomly selected for validation. Due to limited computational power, the large images were cropped into non-overlapping blocks of 1024 × 1024 pixels. The resulting dataset consisted of 615, 252, and 504 labeled images for training, validation, and testing, respectively.

Table 1 presents the results of experiments comparing our proposed method with other methods on the ISPRS Potsdam dataset. The table compares a fully supervised method (SupOnly) with four other state-of-the-art semi-supervised learning methods, namely Mean Teacher [8], CPS [9], U2PL [10], and FixMatch. The evaluation metric used for each method is mIoU. The proposed method demonstrates superior performance compared to other semi-supervised methods at all ratios. It even outperforms fully supervised methods at ratios of 1/4 and below, highlighting its effectiveness when labeled data is scarce. Notably, when using only 1/16 of

Table 1 Comparison of the accuracy of different methods on the Potsdam dataset

Method	1/16(39)	1/8(77)	1/4(154)	1/2(308)	FULL(615)
SupOnly	61.94	65.48	71.12	72.35	73.46
MT	61.86	63.87	67.52	71.07	–
CPS	62.72	65.31	69.94	71.13	–
U2PL	65.43	68.13	70.51	71.38	–
FixMatch	65.88	68.35	70.56	71.45	–
CDPMatch	66.21	68.69	70.77	71.86	–

labeled data, CDPMatch's accuracy significantly improves from 61.86%, 62.72%, and 61.94% (Mean Teacher, CPS, and SupOnly) to 66.21%. When the proportion of labeled data reaches 1/4 or higher, the performance improvement of our proposed semi-supervised framework is not as significant as before. However, CDPMatch still achieves the best performance.

4 Conclusion

This paper introduces CDPMatch, a semi-supervised learning approach for semantic segmentation of remote sensing images based on FixMatch. To prevent perturbation from being limited to the image level and to extend the original perturbation space, we propose a framework that combines image-level and feature-level perturbation (IFPerb). This framework utilizes image-level and feature-level perturbations to complement each other and achieve optimal performance for the semantic segmentation task. To achieve the detection of a predefined perturbation space, we propose a dual-stream perturbation strategy (DBPerb). This strategy uses a single weak-vision image to guide two strong-vision images, taking advantage of contrastive learning for better discriminative characterization. The model achieves equivalent prediction of unlabeled images at both the image and feature levels, ensuring consistency at different levels and strong generalization capability of the model. It has been shown to have significant benefits for the task of semantically segmenting remote sensing images, even when only a small number of labeled images are available.

References

1. Saxena, D., Cao, J.: Generative adversarial networks (GANs) challenges, solutions, and future directions. ACM Comput. Surv. (CSUR) **54**(3), 1–42 (2021)
2. Yang, X., Song, Z., King, I., Xu, Z.: A survey on deep semi-supervised learning. IEEE Trans. Knowl. Data Eng. **35**(9), 8934–8954 (2022)
3. Zhang, H., Zhang, Z., Odena, A., Lee, H.: Consistency regularization for generative adversarial networks (2019). arXiv preprint arXiv:1910.12027

4. Sohn, K., Berthelot, D., Carlini, N., Zhang, Z., Zhang, H., Raffel, C.A., Cubuk, E.D., Kurakin, A., Li, C.L.: Fixmatch: simplifying semi-supervised learning with consistency and confidence. Adv. Neural Inf. Process. Syst. **33**, 596–608 (2020)
5. Caron, M., Misra, I., Mairal, J., Goyal, P., Bojanowski, P., Joulin, A.: Unsupervised learning of visual features by contrasting cluster assignments. Adv. Neural. Inf. Process. Syst. **33**, 9912–9924 (2020)
6. Berthelot, D., Carlini, N., Cubuk, E.D., Kurakin, A., Sohn, K., Zhang, H., Raffel, C.: Remixmatch: semi-supervised learning with distribution alignment and augmentation anchoring (2019). arXiv preprint arXiv:1911.09785
7. Oord, A.V.D., Li, Y., Vinyals, O.: Representation learning with contrastive predictive coding (2018). arXiv preprint arXiv:1807.03748
8. Tarvainen, A., Valpola, H.: Mean teachers are better role models: weight-averaged consistency targets improve semi-supervised deep learning results. Adv. Neural Inf. Process. Syst. **30** (2017)
9. Chen, X., Yuan, Y., Zeng, G., Wang, J.: Semi-supervised semantic segmentation with cross pseudo supervision. In: Proceedings of the IEEE/CVF Conference on Computer Vision and Pattern Recognition, pp. 2613–2622 (2021)
10. Wang, Y., Wang, H., Shen, Y., Fei, J., Li, W., Jin, G., Wu, L., Zhao, R., Le, X.: Semi-supervised semantic segmentation using unreliable pseudo-labels. In: Proceedings of the IEEE/CVF Conference on Computer Vision and Pattern Recognition, pp. 4248–4257 (2022)

Far-Field Speaker Verification Based on Adaptive Feature Alignment

Lingyun Xiang, Jinghan Zhou, and Chengfu Ou

Abstract Due to domain mismatch in far-field environments, far-field automatic speaker verification (ASV) shows inferior performance. To address this issue, we propose a far-field ASV method based on adaptive feature alignment (AFA-FSV). Initially, this method employs RepVGG model to extract general speaker features separately from both the source and target domains. To accurately capture the overall distributions of the source and target domains while further considering the paired distances of the same speaker features between the two different domains, a speaker embedding domain distance metric method is designed. It adaptively allocates different weights to generic speaker features based on the magnitude of distribution differences between the source and target domains, thereby aligning the deep features of the same speaker across domains to achieve domain-invariant feature extraction. Finally, precise speaker classification is performed based on the extracted domain-invariant features. Experimental results on the HI-MIA dataset demonstrate that AFA-FSV outperforms other competitive domain-adaptive methods.

Keywords Far-field speaker verification · Domain-invariant · Knowledge transfer

1 Introduction

Due to cross-domain challenges in far-field environments, ASV systems often experience a sharp decline in verification accuracy under far-field conditions [1].

L. Xiang (✉)
School of Computer and Communication Engineering, Changsha University of Science and Technology, Changsha, China
e-mail: xiangly@csust.edu.cn

J. Zhou
College of Digital Information, Shanghai Sipo Polytechnic, Shanghai, China
e-mail: zhoujh@sipo.sicfl.edu.cn

C. Ou
College of Cyber Security, Jinan University, Guangzhou, China
e-mail: hahally@stu2025.jnu.edu.cn

J. S. Park et al. (eds.), *Advanced Multimedia and Ubiquitous Engineering*, Lecture Notes in Electrical Engineering 1475, https://doi.org/10.1007/978-981-95-1565-3_14

To alleviate domain mismatch issues and enhance far-field verification performance, domain adaptation techniques have been introduced into ASV methods [2]. Early approaches directly applied domain adaptation to the backend of x-vector systems with Probabilistic Linear Discriminant Analysis [3]. Subsequent research has focused on designing and training domain-invariant speaker embeddings to alleviate the problem of degraded speaker verification accuracy caused by domain mismatch between training and testing datasets. Furthermore, metric learning [4] has also been used in ASV. Contrastive loss [5] and triplet loss [6] have been used to optimize distance metrics.

While the methods mentioned address the domain mismatch issue, they mainly focus on capturing the overall distribution of two domains, neglecting the paired distances of the same speaker features between them. Ideally, the distance distribution between samples of the same speaker should be similar in their respective feature spaces. To tackle this, we propose AFA-FSV. It involves general speaker feature extraction, domain-invariant feature extraction, and speaker classification. In general speaker feature extraction phase, RepVGG model is used to extract general speaker features from both the source domain (close-talking utterances) and the target domain (close-talking and far-field utterances). During domain-invariant feature extraction, we design a speaker embedding domain distance metric method (SEDDM) for far-field ASV, which incorporates speaker embedding feature information metric and correlation alignment metric [7] to capture the overall distribution of both source and target domains. Guided by the proposed SEDDM, weights are adaptively assigned to general speaker features of different domains, aligning deep features of the same speaker between domains to achieve domain-invariant feature extraction. Finally, in the speaker classification phase, domain-invariant features are processed through fully connected (FC) layers and activation functions to determine speaker probability distributions, minimizing cross-entropy loss for speaker identification. Experimental results on the HI-MIA dataset show significant improvements of AFA-FSV over several competing methods.

2 AFA-FSV

The overall framework of our proposed far-field speaker verification method, AFA-FSV, is illustrated in Fig. 1. The orange paths in the diagram represent the data flow direction from the source domain, while the green paths represent the data flow direction from the target domain. Each phase will be described in detail below.

2.1 General Speaker Feature Extraction

To comprehensively capture the features of speakers, we use RepVGG model [8] to extract general speaker features from both the source and target domains' speech

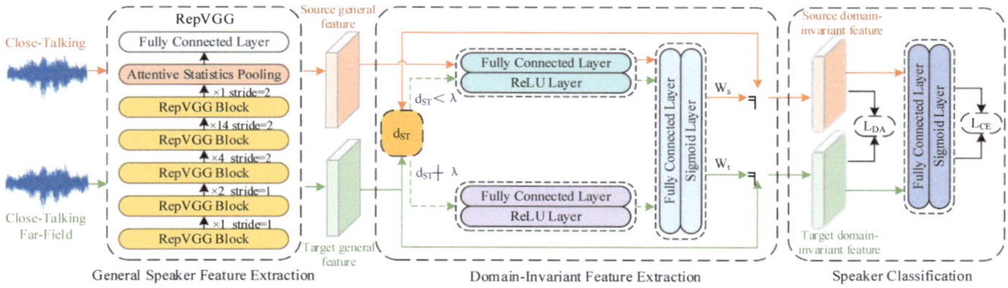

Fig. 1 The overall framework of our proposed model

signals. This model consists of a RepVGG-A backbone [9], an attentional statistics pooling (ASP) layer [10], and a FC layer. Given input speech $x_s \in \mathbf{X}_s$ and $x_t \in \mathbf{X}_t$, the general speaker feature representations extracted from both the source $\mathbf{H}_s \in \mathbb{R}^{1\times256}$ and target domains $\mathbf{H}_t \in \mathbb{R}^{1\times256}$ using the RepVGG model.

2.2 *Domain-Invariant Feature Extraction*

To address domain mismatch in far-field ASV, we propose SEDDM. It captures overall distributions of both source and target domains while considering the paired distances of general speaker features between them. Then, we perform domain-invariant feature extraction using SEDDM, adaptively assigning weights based on domain distribution differences, aligning speaker deep features from both domains. We detail SEDDM and then describe domain-adaptive weights based on it.

Speaker Embedding Domain Distance Metric (SEDDM). SEDDM is composed of two components: the speaker embedding feature information metric d_f and the CORAL metric d_c. Specifically, it is described as follows:

(1) d_f: Inspired by [11], d_f is to transfer robust features extracted from the source domain to the target domain. Specifically, we consider speaker feature vectors from the source domain as anchor points. Then we consider the samples from the target domain with the same label as the anchor point as positive samples, while those with different labels as negative samples. Domain adaptation is achieved by reducing the distance between positive sample pairs and increasing the distance between negative sample pairs. The expression of d_f is as follows:

$$d_f(\mathbf{H}_s, \mathbf{H}_t) = -\frac{1}{N}\sum_{i=1}^{N}\log\frac{\exp(\langle \mathbf{H}_s, \mathbf{H}_t\rangle)}{\sum_{\alpha\in A(i)}\exp(\langle \mathbf{H}_s, \mathbf{H}_\alpha\rangle)} \tag{1}$$

where, $\langle\cdot\rangle$ represents the inner product of feature vectors. $\mathbf{H}_s$ and $\mathbf{H}_t$ have the same speaker label, while $\mathbf{H}_s$ and $\mathbf{H}_\alpha$ have different labels. $A(i)$ is the index of the training samples from the source domain, and its corresponding sample has a speaker label inconsistent with $\mathbf{H}_s$.

(1) d_c: It uses the second-order moment to estimate the distribution of samples, and taking the difference as the distance between the source domain and the target domain, this can represent the dispersion of features. The expression of d_c is as follows:

$$d_c(\mathbf{H}_s, \mathbf{H}_t) = \frac{1}{4d^2}\|C_s - C_t\|_F^2 \tag{2}$$

where, $\|\cdot\|_F$ represents the Frobenius norm of a matrix, C_s and C_t denote the feature covariance matrices of the source and target domains respectively, and d represents the dimensionality of the features.

(2) d_{SEDDM}: d_f can accurately measure the overall feature distribution difference between two domains, while d_c achieves domain adaptation by making the feature representations of the same speaker more consistent across the two domains. Therefore, we combine the strengths of d_f and d_c, selecting d_f as the primary component of SEDDM and d_c as a controllable variable to further align the deep features of the same speaker. Thus, a metric function for the embedding domain distance d_{SEDDM} between the speaker in the source domain and the target domain is designed. The expression of d_{SEDDM} is as follows, γ is a hyperparameter, and we set its value to 10.

$$d_{SEDDM}(\mathbf{H}_s, \mathbf{H}_t) = d_f(\mathbf{H}_s, \mathbf{H}_t) \cdot e^{(\gamma \cdot d_c(\mathbf{H}_s, \mathbf{H}_t))} \tag{3}$$

SEDDM-based Domain-Adaptive Weight. We design a segmentation structure where domain weights $\mathbf{W}_t$ and $\mathbf{W}_s$ indicate the relevance of features in the source and target domains respectively. If the distance between domains is small, suggesting similarity, the target and source domains can share the same branch and domain weights; otherwise, independent branches should be used to derive distinct domain weights. This can be expressed as:

$$\mathbf{W}_s = \sigma(FC(FC_s(\mathbf{H}_s))) \tag{4}$$

$$\mathbf{W}_t = \begin{cases} \sigma(FC(FC_s(\mathbf{H}_t))), \textit{ if } d_{ST} < \lambda \\ \sigma(FC(FC_t(\mathbf{H}_t))), \text{ otherwise} \end{cases} \tag{5}$$

where, $d_{ST} = d_{SEDDM}(\mathbf{H}_s, \mathbf{H}_t)$ represents the distribution difference between the source and target domain features. $FC_s(\cdot)$ and $FC_t(\cdot)$ denote the FC layers with ReLU activation function in the source and target domains respectively. $FC(\cdot)$ represents the FC layer on the rightmost side of the Fig. 1. $\sigma(\cdot)$ represents the sigmoid activation function. Then we obtain weighted features from the source domain and the target domain ($\mathbf{H}_s'$ and $\mathbf{H}_t'$). Consequently, it enables robust domain-invariant feature representation under different domain discrepancies. We update the network by minimizing the distribution distance between samples from the source and target domains. For each pair of source and target domains, the domain adaptation loss is expressed as:

$$L_{DA} = d_{SEDDM}\left(\mathbf{H}_s', \mathbf{H}_t'\right) \tag{6}$$

2.3 Speaker Classification

The source domain's classification loss uses the cross-entropy loss function L_{CE}. The total loss in model training is defined as:

$$L_{total} = \lambda_1 * L_{CE} + \lambda_2 * L_{DA} \tag{7}$$

where λ_1 and λ_2 are hyperparameters adjusting the weights of the two losses.

3 Experiments

3.1 Experimental Setup

We compare our methods with several domain-adaptive methods, namely cosine distance [10], MLT [11], KL divergence [12] and MMD [13] loss. All experiments were conducted on the HI-MIA dataset [14]. All methods employ ASP and use 200 frames of 80-dimensional Fbank features as input. The classifier uses AAM-Softmax. The sampling rate was set to 16kHz. During the testing phase, cosine similarity was used to compute scores between two speakers' utterances, which were used to calculate evaluation metrics such as EER and minDCF.

3.2 Experimental Results and Analysis

Comparison with domain-adaptive methods. To validate the accuracy of our proposed method, we conduct relevant experiments as shown in Table 1. Compared with other methods, AFA-FSV demonstrates the best speaker verification performance.

Ablation Study. To further validate the effectiveness of our model, we conduct ablation experiments. "wo-AW" denotes "without adaptive weights", "wo-SEDDM" denotes "without SEDDM" and "wo-λ" denotes "without weight assignment strategy", where the general speaker features of the source and target domains are used

Table 1 Comparison with domain-adaptive methods

Methods	EER	minDCF
Cosine distance [10]	4.62	0.4168
MLT [11]	7.52	0.6350
KL divergence [12]	5.16	0.4644
MMD [13]	7.00	0.5820
AFA-FSV (Ours)	**3.98**	**0.3886**

Table 2 Ablation study on AFA-FSV

Methods	EER	minDCF
wo-AW	4.46	0.5020
wo-SEDDM	4.80	0.4618
wo-λ	4.30	0.3986
AFA-FSV(Ours)	**3.98**	**0.3886**

in the same FC layer to calculate corresponding weights. From Table 2, we observe that our method achieved better performance.

4 Conclusion

This paper addresses the domain mismatch issue by proposing a far-field ASV method called AFA-FSV. We design a Speaker Embedding Domain Distance Metric (SEDDM), which not only captures the overall distribution differences between the source and target domains but also aligns the deep features of the same speaker between these domains. Domain-invariant feature extraction based on SEDDM adaptively assigns weights to general speaker features from different domains. Experimental results demonstrate that SEDDM proposed in this paper is a more effective metric for matching features between the source and target domains in existing domain-adaptive far-field ASV methods. Moreover, the performance of AFA-FSV exceeds that of other competitive domain-adaptive methods.

References

1. Li, L., et al.: A principle solution for enroll-test mismatch in speaker recognition. IEEE/ACM Trans. Audio Speech Language Process. **30**, 443–455 (2022)
2. Pan, S.J., Yang, Q.: A survey on transfer learning. IEEE Trans. Knowl. Data Eng. **22**(10), 1345–1359 (2009)
3. Prince, S.J.D., Elder, J.H.: Probabilistic linear discriminant analysis for inferences about identity. In: 11th International Conference on Computer Vision, pp. 1–8. IEEE (2007)

4. Bai, Z., Zhang, X.-L., Chen, J.: Speaker verification by partial AUC optimization with mahalanobis distance metric learning. IEEE/ACM Trans. Audio Speech Language Process. **28**, 1533–1548 (2020)
5. Rahman Chowdhury, F.A.R., et al.: Attention-based models for text-dependent speaker verification. In: ICASSP, pp. 5359–5363. IEEE (2018)
6. Zhang, C., Koishida, K., Hansen, J.H.L.: Text-independent speaker verification based on triplet convolutional neural network embeddings. IEEE/ACM Trans. Audio Speech Language Process. **26**(9), 1633–1644 (2018)
7. Sun, B., Saenko, K: Deep coral: correlation alignment for deep domain adaptation. In: Computer Vision–ECCV 2016 Workshops, pp. 443–450. Springer International Publishing, Amsterdam (2016)
8. Zhao, M., Ma, Y., Liu, M., Xu, M.: The Speakin System for Voxceleb Speaker Recognition Challenge 2021. arXiv preprint arXiv:2109.01989 (2021)
9. Ma, Y., Zhao, M., Ding, Y., Zheng, Y., Liu, M., Xu, M.: Rep Works in Speaker Verification. arXiv preprint arXiv:2110.09720 (2021)
10. Jung, J.-W., Heo, H.-S., Shim, H.-J., Yu, H.-J.,: Short utterance compensation in speaker verification via cosine-based teacher-student learning of speaker embeddings. In: Automatic Speech Recognition and Understanding Workshop, pp. 335–341. IEEE (2019)
11. Zhang, L., Wang, Q., Lee, K.A., Xie, L., et al.: Multi-Level Transfer Learning from Near-Field to Far-Field Speaker Verification. arXiv preprint arXiv:2106.09320 (2021)
12. Heo, H.-S., Jung, J.-W., Shim, H.-J., Yu, H.-J.: Acoustic Scene Classification Using Teacher-Student Learning with Soft-labels. arXiv preprint arXiv:1904.10135 (2019)
13. Lin, W., et al.: Multi-level deep neural network adaptation for speaker verification using MMD and consistency regularization. In: ICASSP, pp. 6839–6843. IEEE (2020)
14. Qin, X., Bu, H., Li, M.: Hi-mia: a far-field text-dependent speaker verification database and the baselines. In: ICASSP, pp. 7609–7613. IEEE (2020)

A Data Skew Greedy Optimization Strategy in Spark Heterogeneous Clusters

Chen Huang and Xiaoyong Tang

Abstract In distributed computing frameworks, data skew is one of the main factors affecting performance improvement. The essence of this problem is that there are differences in the computing load of partition tasks, resulting in node load imbalance, especially in heterogeneous scenarios. This paper proposes a data skew greedy optimization strategy (HSDSGO) based on heterogeneous Spark to solve the data skew problem, which comprehensively considers differences in node performance and data locality. Firstly, we use an improved reservoir sampling algorithm to sample the input data in the Map phase, obtaining the key distribution in the entire dataset. Secondly, the capacity size of mapping partition tasks is determined based on the evaluation of node computational capabilities. Finally, the data is partitioned using a best-fit decreasing algorithm. Experimental results demonstrate that the HSDSGO algorithm not only reduces task execution time but also maximizes the utilization of cluster resources.

Keywords Data skew · Heterogeneous · Data locality

1 Introduction

Spark jobs mainly consist of two stages: the Map stage and the Reduce phase. The shuffle phase acts as an intermediary and is responsible for maintaining the read/write relationship between them. However, when the data distribution is unbalanced, it will lead to uneven load between Map tasks and Reduce tasks. This phenomenon is often called data skew in distributed computing frameworks [1].

In order to alleviate the impact of data skewness on system performance, previous researchers have made optimizations in the following aspects: partition data balancing [2, 3], reducing shuffle processes [4, 5], improving data locality [6–8], and improving resource utilization [9–11]. Although the above research has to

C. Huang (✉) · X. Tang
Changsha University of Science and Technology, Changsha, China
e-mail: 21208051575@stu.csust.edu.cn

J. S. Park et al. (eds.), *Advanced Multimedia and Ubiquitous Engineering*, Lecture Notes in Electrical Engineering 1475, https://doi.org/10.1007/978-981-95-1565-3_15

some extent alleviated data skewness, it has not considered how to solve data skewness in heterogeneous scenarios. This is because in heterogeneous scenarios, even if the amount of partition tasks is equal, the execution time of partition tasks will be extended due to differences in node capabilities.

Inspired by the above research on optimizing the data skew problem, this paper comprehensively considers the performance differences between nodes and the locality of data in heterogeneous scenarios, and proposes a data skew greedy optimization algorithm based on heterogeneous Spark. The main contributions of this paper are as follows:

- An improved reservoir sampling algorithm is proposed to sample the input data in the Map stage and predict the distribution of key-value pairs in the partition.
- An algorithm for evaluating node computing capabilities is proposed. The algorithm determines mapping tasks by node computing power and data locality.
- An improved best-fit algorithm is proposed to allocate data to reduce data skew.

2 Heterogeneous System Architecture

The framework of this article is shown in Fig. 1, we propose a data skew greedy optimization strategy based on heterogeneous Spark to solve the shortcomings of partitioning strategy. The algorithm mainly includes three parts: improved reservoir sampling prediction, node weight evaluation algorithm, and improved greedy data partitioning algorithm.

3 Data Skew Greedy Optimization Strategy

3.1 Improved Reservoir Sampling Prediction

Improved reservoir sampling prediction mainly uses an improved reservoir sampling algorithm to partially sample the input dataset in the map stage, predict the distribution of keys in the dataset, and generate corresponding data partition strategies. This sampling method not only allows for random sampling without knowing the size of the dataset, but also predicts accurate data distribution at a lower cost.

We have verified through experiments that when the sampling rate is 15%, the cost is lower, and the distribution of the sampled data is more accurate.

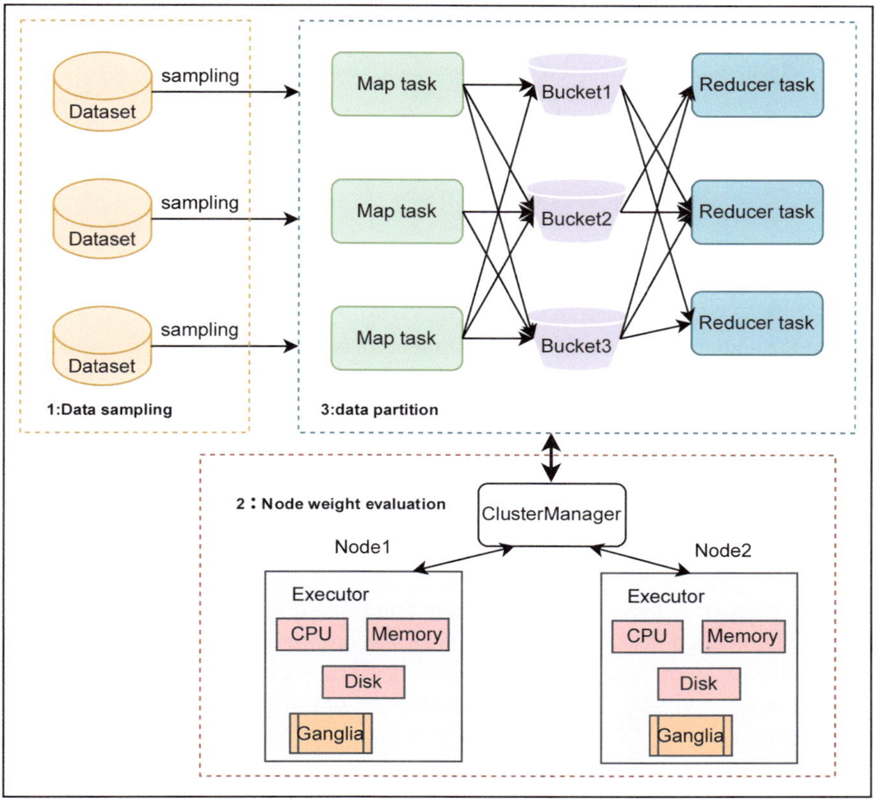

Fig. 1 HSDSGO framework

3.2 *Node Weight Evaluation Algorithm*

Node weight evaluation algorithm mainly involves monitoring the resource usage of nodes in the cluster and evaluating the nodes using the analytic hierarchy process to obtain corresponding weight values. We map nodes to tasks by considering weight size and data locality. For illustrative purposes, Table 1 describes some important variable declarations.

Spark consists of n cluster nodes, and the computing power of the nodes is C_i. The input dataset size for the Map stage is $\mathrm{D} = \{k_1, k_2, \ldots k_{\mathrm{j}}, \ldots, k_m\}$. The weight of node i can be expressed as:

$$f_i = \frac{C_i}{\sum_{i=1}^{n} C_i} \tag{1}$$

The average execution time T_e of the entire Spark stage can be expressed as:

Table 1 Node hardware configuration

Host name	CPU frequency	CPU core	Memory	Disk
Master	IntelXeon@2.5GHZ	4	16G	50G
Slave1	IntelXeon@2.5GHZ	2	8G	30G
Slave2	IntelXeon@2.5GHZ	4	32G	60G
Slave3	IntelXeon@2.2GHZ	8	64G	80G

$$T_e = \frac{D}{\sum_{i=1}^{n} C_i} \tag{2}$$

And the actual execution time of partition tasks is T_i. In heterogeneous situations, if the actual execution time is close to the average execution time, the following mapping matrix Q can be represented as:

$$Q = \begin{bmatrix} q_{11} & \cdots & q_{1n} \\ \vdots & q_{ij} & \vdots \\ q_{m1} & \cdots & q_{mn} \end{bmatrix}, 1 \le i \le m, 1 \le j \le n \tag{3}$$

where q_{ij} represents the computational cost that task i allocates to node j to execute the task, and if this data is close to T_e, then the optimal adaptation node of task i is j, otherwise the optimal node is re-founded. After obtaining the mapping matrix between nodes and tasks, the threshold for partitioning tasks can be calculated as follows:

$$d^i_{\max} = \sum_{j=1}^{m} k_j \cdot f_i \tag{4}$$

where $d^i_{\max}$ indicates the capacity of the partition task, and the larger the capacity, the stronger the computing power of the node.

3.3 Improved Greedy Data Partition Algorithm

Improved greedy data partition algorithm is to distribute the intermediate data of the Map stage to different reduce tasks according to the improved optimal adaptive descending algorithm. The main goal of the data partition algorithm is to allocate the key cluster to appropriate partitions according to the size of the partition capacity, so as to minimize the job execution time.

First, the dataset is assigned to the reduce task according to the best adaptation descending algorithm until all datasets are allocated. The key distribution D and the computing power of nodes C_i are taken as inputs in algorithm 1, determine the partition capacity d^i_{max}. First, the keys are sorted in descending order, and the capacity of the partition task is also sorted in descending order, starting with the largest key. In general, the largest key can be allocated to the maximum partition task. Select $d^i_{remain} \geq k_j$ partitions with a capacity close to k_j from all Reduce tasks and allocate them until all keys are allocated.

Algorithm 1: Data partition algorithm

Input: Key distribution D, computing power of the nodes C_i

Output: The reduce partition index corresponding to the key

$SortKey();$

$d^i_{\max} = \sum_{j=1}^{m} k_j \cdot f_i;$

$Sortd_{\max}();$

$\boldsymbol{for}(j = 1; j <= m; j++)\boldsymbol{do}$

  $\boldsymbol{for}\ (i = 1; i <= n; i++)\boldsymbol{do}$

    $\boldsymbol{if}(d^i_{\max} \geq k_j)\boldsymbol{then}$

      $\boldsymbol{if}\ d^i_{\max} - k_j \leq \min(d^i_{\max} - k_j)\ \boldsymbol{then}$

        $\mathrm{P} = PlacementKeyToReduce(i);$

$UpdateAndSortReduce(i);$

endif

endif

endfor

endfor

return*P*

In data partition algorithms, data allocation is based on the computing power and data locality of each node, which can maximize the utilization of node performance and calculate more data. Overall, the time complexity of the data partitioning algorithm is $O(m \times \log^n)$, which is due to filtering through quick sorting when selecting tasks.

4 Experiment and Result Analysis

4.1 Experiment Setting

The Spark cluster environment consists of 4 nodes (1 master node and 3 slave nodes), and the cluster resource manager is Standalone. To reflect the heterogeneity of the cluster environment, the resource configuration of each node is different, where the master, Slave1, and Slave2 are deployed on the same server, and Slave3 is deployed on another server. Table 1 lists the hardware configurations of the nodes.

In order to evaluate the performance of this algorithm, we selected two representative benchmark programs: WordCount, and selected task execution time as the indicator for performance evaluation. Task execution time refers to the time from the start to the end of the job run. Since we want to evaluate the performance of the HSDSGO algorithm under different tilt degrees, the data sets used in this article are generated by Zipf [12] with customized data skewness. The sizes of the data sets are 2 GB, 4 GB and 8 GB respectively. In this experiment, we selected the following three partitioning strategies for comparison:

DEFH algorithm (Hash Partition) is Spark's default partitioning strategy. It works by evenly spreading data across multiple partitions based on its hash value.

Range algorithm is also a commonly used partitioning strategy in Spark. It divides data into different partitions based on the range of the data.

The SCID [13] algorithm is a data splitting-merging algorithm. This algorithm first predicts the size of the data through a sampling algorithm, finds skewed key-value pairs, and splits the severely skewed Keys through continuous iterations and assigns them to different Reduce tasks, ultimately balancing the load of the data in the Reduce stage.

4.2 Performance Evaluation

To verify the effectiveness of our algorithm in heterogeneous environments, we ran the WordCount program using the groupByKey operator. We generated custom skewed datasets using Zipf and tested the performance differences between our algorithm and the other three partitioning algorithms at different scales.

Figure 2 shows the variation of task execution time for HSDSGO and three other partitioning algorithms in WordCount under different degrees of skew. From Fig. 1 (a), when the degree of data skew is not significant, the execution time of DEFH is the shortest. As the data scale continues to expand and the skewness increases from 0.2 to 1, the execution of all four partitioning algorithms has increased, with HSDSGO showing a slower growth rate in task execution time. In Fig. 1c, when the input dataset size is 8 GB and the skewness reaches 0.8, the task execution time of HSDSGO is reduced by 25.1% compared to the DEFH algorithm.

5 Conclusion

Our work is mainly divided into three aspects: (1) In the Map stage, we designed an improved reservoir sampling algorithm to predict the input dataset. (2) We consider the computing power and data locality of nodes and establish a mapping matrix

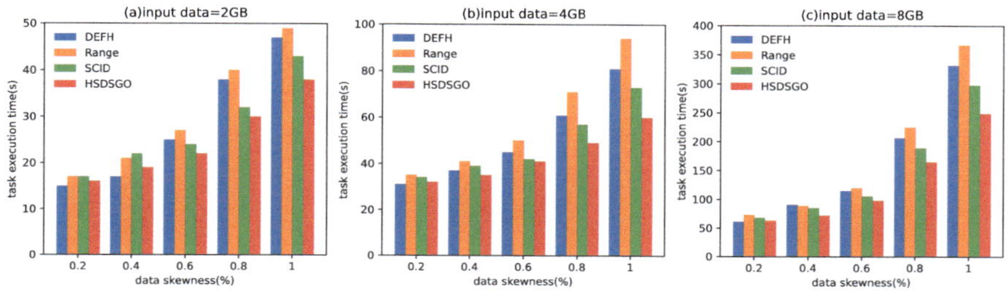

Fig. 2 The impact of data skewness on task execution time in WordCount with different input data sizes

between nodes and tasks, and then set the capacity size of the corresponding partition tasks for nodes. (3) Aiming at the problem of imbalanced data distribution in heterogeneous environments, an improved best fit descending algorithm is proposed to partition data. The experimental results show that our proposed HSDSGO algorithm can effectively alleviate the situation of unreasonable data allocation in heterogeneous scenarios, reduce the overall running time of Spark jobs, and improve the resource utilization of the cluster.

References

1. Chen, Q., Yao, J., Xiao, Z.: Libra: lightweight data skew mitigation in MapReduce. IEEE Trans. Parallel Distrib. Syst. **26**(9), 2520–2533 (2014)
2. Li, C., Zhang, Y., Luo, Y.: Intermediate data placement and cache replacement strategy under Spark platform. J. Parallel Distrib. Comput. **163**, 114–135 (2022)
3. Tang, Z., Lv, W., Li, K., et al.: An intermediate data partition algorithm for skew mitigation in spark computing environment. IEEE Trans. Cloud Comput. **9**(2), 461–474 (2018)
4. Monica, V., Parambalath, N.K.: Shuffle phase optimization in spark. In: 2017 International Conference on Advances in Computing, Communications and Informatics (ICACCI), pp. 1028–1034. IEEE (2017)
5. Raj, S., Ramesh, D., Sethi, K.K.: A spark-based Apriori algorithm with reduced shuffle overhead. J. Supercomput. **77**(1), 133–151 (2021)
6. Fu, Z., He, M., Tang, Z., Zhang, Y.: Optimizing data locality by executor allocation in spark computing environment. Comput. Sci. Inf. Syst. **20**(1), 491–512 (2023)
7. Guo, W., Huang, C., Tian, W.: Handling data skew at reduce stage in spark by reduce partition. Concurr. Comput.: Pract. Exp. **32**(9): e5637 (2020)
8. Fu, Z., He, M., Tang, Z., et al.: Optimizing data locality by executor allocation in reduce stage for spark framework. In: International Conference on Parallel and Distributed Computing: Applications and Technologies. Springer International Publishing, Cham, pp. 349–357 (2021)
9. Xiujin, S., Yueqin, Q.: An algorithm of data skew in spark based on partition. In: 2020 International Conference on Computers, Information Processing and Advanced Education (CIPAE), pp. 217–222. IEEE (2020)
10. Pan, F., Xiong, J., Shen, Y., et al.: H-scheduler: storage-aware task scheduling for heterogeneous-storage spark clusters. In: 2018 IEEE 24th International Conference on Parallel and Distributed Systems (ICPADS), pp.1–9. IEEE (2018)
11. Xu, L., Butt, A.R., Lim, S.H., et al.: A heterogeneity-aware task scheduler for spark. In: 2018 IEEE International Conference on Cluster Computing (CLUSTER), pp. 245–256. IEEE (2018)

12. Axtell, R.L.: Zipf distribution of US firm sizes. Science **293**(5536), 1818–1820 (2001)
13. Tang, Z., Zhang, X., Li, K., et al.: An intermediate data placement algorithm for load balancing in spark computing environment. Futur. Gener. Comput. Syst. **78**, 287–301 (2018)

A Method for Invalid Wind Power Data Identification Based on Segmented Quartiles and Peak Detection

Junsheng Chen, Zhongheng Zhang, Benchuan Li, Mingjie Liu, and Changhao Piao

Abstract Due to various factors such as communication interference, operational faults, maintenance shutdowns, protective shutdowns, and grid dispatching instructions, wind farms often accumulate a significant amount of invalid data in their collected operational data. This low-quality data presents challenges for tasks such as wind power prediction and wind turbines condition monitoring. To address this issue, this paper presents a novel method for identifying invalid data in wind turbine operations, which combines segmented quartiles and curve morphological feature detection. The proposed method first employs the segmented quartiles approach to identify sparse invalid data points. Subsequently, leveraging the curve morphological characteristics associated with downrating invalid data, a peak detection algorithm integrated with curve fitting techniques is utilized to identify invalid data points specifically related to downrating operations. To validate the effectiveness and applicability of the proposed method, both real-world wind farm data and artificially generated simulation data are employed. The results demonstrate the efficacy and universality of the proposed approach in identifying and filtering out invalid data, thereby enhancing the reliability of wind power prediction and wind turbines condition monitoring.

Keywords Wind power data · Peak detection · Invalid data identification · Wind turbine

1 Introduction

The operational monitoring data of wind turbine is fundamental for anomaly detection, fault diagnosis, health management, and wind power forecasting [1]. However, due to the volatility, intermittency, and randomness of wind energy, the raw collected operational data, especially wind power data, contains a high proportion of invalid

J. Chen (✉) · Z. Zhang · B. Li · M. Liu · C. Piao
College of Automation, Chongqing University of Posts and Telecommunications, Chongqing, China
e-mail: chenjunsheng@cqupt.edu.cn

J. S. Park et al. (eds.), *Advanced Multimedia and Ubiquitous Engineering*, Lecture Notes in Electrical Engineering 1475, https://doi.org/10.1007/978-981-95-1565-3_16

data. This makes it difficult to use directly for subsequent wind power forecasting and data modeling research [2]. Therefore, effectively identifying invalid data in raw wind power data is essential for accurately perceiving the operating status of wind turbines and making informed wind power-related decisions.

Traditional methods for handling invalid wind power data primarily fall into two categories: machine learning-based approaches and statistical analysis-based methods. Yan et al. [3] employed the DBSCAN clustering algorithm to identify invalid wind power data. Sun et al. [4] leverages the interconnectedness of data similarity to address the invalid wind power data points effectively. However, a significant limitation lies in the requirement to establish neighboring models for multiple wind turbines, leading to a considerable workload. Zou et al. [5] introduced a method for identifying invalid wind power data using Thompson tau and the quartile approach. However, this method exhibits a relatively high misrecognition rate for normal data, potentially reducing data utilization efficiency. Additionally, image analysis-based methods have also been explored for identifying invalid wind power data. Long et al. [6] employed image edge detection technology to detect and remove sparse invalid wind power data effectively. Liang et al. [7] pursued a distinct approach by transforming the scatter plot of wind power into a digital image, enabling the identification of invalid data through minimization of the image threshold. Nevertheless, adjusting the grayscale threshold parameter is necessary to accommodate varying wind power datasets. Furthermore, it might struggle to accurately distinguish erroneously flagged invalid data, especially in proximity to the cut-out wind speed.

In summary, the primary challenges in identifying invalid wind power data emerge in the following aspects: (a) methods necessitate manual adjustment of parameters based on data characteristics; (b) Probabilistic and statistical methods entail extensive data calculations, resulting in prolonged processing time; (c) Image analysis-based methods depend on real-time, accurate, and stable reference data to support analysis and decision-making.

To overcome these limitations, this manuscript introduces an innovative method for identifying invalid data by integrating segmented quantile analysis with peak feature detection. By examining the underlying causes and distribution patterns of invalid data, the invalid wind power data is categorized into three types, including bottom cumulative invalid data, sparse invalid data, and downrating invalid data. The detailed distribution characteristics of these types are elaborated in [8] and are not reiterated here. Furthermore, we propose specific cleaning methods tailored for each type of invalid data to ensure precise identification.

2 Proposed Methodology

2.1 Bottom-Stacked Invalid Data Identification Based on Rule-Based Filtering

Based on the distribution characteristics of Bottom-stacked invalid data and the operating principles of wind turbines, the identification of such invalid data can be achieved through the application of rule-based screening. This entails flagging and cleansing data points where the power value is less than or equal to zero, the wind speed is below the cut-in wind speed, or the wind speed exceeds the cut-out wind speed.

2.2 Sparse Invalid Data Identification Based on Segmented Quartile Method

In this manuscript, the wind power data, post-filtering via rule-based screening (denoted as $\mathbf{D} = [d_1, d_2, d_3, \ldots, d_n]$), is divided into K equally spaced intervals along the power axis. Subsequently, the quartile method is employed to process the wind power data within each interval. Sparse invalid data are rectified to obtain a non-sparse wind power dataset (denoted as $\mathbf{W}$). The quartile method demonstrates superior efficacy in identifying sparse data, involving the determination of the lower quartile (Q_L) and upper quartile (Q_U) using the approach outlined in [5], followed by the calculation of the interquartile range (IQR):

$$IQR = Q_U - Q_L \tag{1}$$

Any data point d_i ($i = 1, 2, 3, \ldots, n$) that that fails to satisfy the specified condition ($Q_L - 1.5IQR \leq d_i \leq Q_U + 1.5IQR$) is identified as sparse invalid data. Following thorough deliberation, the value of K is selected to be 150 in this manuscript.

2.3 Downrating Invalid Data Identification Based on Wave Peak Detection

a) **Wind speed width parameter h**

In this paper, the difference between the maximum and minimum wind speeds corresponding to the same power value is defined as the wind speed width at that power, denoted as h and the wind speed width (h) is calculated as:

$$h = Q_U - Q_L + 3IQR \tag{2}$$

The wind speed width parameter and power parameter are determined for each region, yielding the power-wind speed width dataset (denoted as **H**).

b) **Data Smoothing**

To facilitate subsequent processing without altering the inherent fluctuation trends of the data, a Gaussian filter is adopted for smoothing the dataset **H** [9]:

$$smoothed(x) = \sum\nolimits_{j=1}^{m} (w(x_j) \times I(x_j)) \quad (3)$$

$$w(x_j) = \frac{1}{\sigma\sqrt{2\pi}} e^{\frac{-(x_j-x)^2}{2\sigma^2}} \quad (4)$$

where $I(x_j)$ represents the input wind speed width data, and $smoothed(x)$ is the smoothed output wind speed width data, $w(x_j)$ is the weight of x_j. σ represents the standard deviation of the Gaussian operator.

c) **Wave peak detection algorithm**

Due to the characteristics of downrating invalid data, the wind speed width parameter of data $\mathbf{H_1}$ may experience sudden increases at the downrating band, forming peaks. Peak detection essentially involves identifying local maxima in the data, satisfying the following criteria: (1) the data values of the adjacent points are both smaller than the data value at that point, i.e., $h_p > h_{p+1}$, $h_p > h_{p\text{-}1}$; (2) the prominence is relatively high, where the prominence (pm) parameter is defined as the vertical distance between the peak's height (or maximum value) and the lowest value on either side ($h_{\max} - h_{\min}$). This value quantifies the "prominence" of the peak [10].

After identifying the peak's location, it becomes imperative to ascertain the starting power (p_s) and ending power (p_e) of the peak. It should be noted that p_s and p_e meet the following criteria: the slope at p_s in $\mathbf{H_1}$ is zero, or there is a change in the sign of the slope between p_s and $p_{s\text{-}1}$, and the slope at p_e in $\mathbf{H_1}$ is zero, or there is a change in the sign of the slope between p_e and p_{e+1}.

d) **Curve fitting**

This manuscript identifies invalid data in preprocessed wind power data by fitting wind speed width curves and wind power curves [11]. After obtaining p_s and p_e through peak detection, the data within each p_s to p_e range in the smoothed dataset $\mathbf{H_1}$ are masked to obtain the fitted wind power-wind speed width dataset $\mathbf{H_2}$. This paper proposes an exponential curve fitting method for $\mathbf{H_2}$ [12]:

$$h(x) = a_1^{(x-b_1)} + c_1 \quad (5)$$

where a_1, b_1, and c_1 represent the coefficients of the curve, and x denotes power. For non-sparse wind power dataset **W**, the wind power data within each peak from p_s to p_e, as well as the wind power data corresponding to wind speeds exceeding the rated wind speed v_r, are masked to obtain the fitted wind power dataset $\boldsymbol{P_v}$. Based on the introduction of standard power curves in reference [13], this paper selects a standard cubic function to fit $\boldsymbol{P}_v$:

$$v(x) = a_2x^3 + b_2x^2 + c_2x + d_2 \tag{6}$$

where a_2, b_2, c_2, and d_2 represent the coefficients of the curve.

By applying $h(x)$ and $v(x)$ to the dataset $\mathbf{W}$, we obtain $h(p)$ and $v(p)$ corresponding to each power level p, along with the actual wind power data v_p. If the wind speed v_p is less than $v(p) - h(p)$ or greater than $v(p) + h(p)$, the data is considered as downrating invalid data and is subsequently cleaned. For data within the range from the rated wind speed v_r to the cut-out wind speed v_o, only wind power data within $\pm$ 5% of the rated power P_r are retained.

3 Case Study

3.1 Identification and Analysis of Actual Wind Farm Data

To validate the effectiveness and generality of the proposed method in this paper, actual operational data from wind turbines at a wind farm in China are utilized for verification analysis. The wind turbines at this wind farm record data every ten minutes. The rated parameters of the turbines are as follows: a cut-in wind speed of 3 m/s, a rated wind speed of 13 m/s, a cut-out wind speed of 25 m/s, and a rated power of 2000 kW.

After conducting the rule-based screening, the non-sparse wind power dataset $\mathbf{W}$ is acquired by eliminating invalid data using the segmented quartile method. The sparse invalid data identification results of turbines 1 and 8 are depicted in Fig. 1.

A comparison between the smoothing results and the corresponding non-sparse invalid dataset $\mathbf{W}$ is presented in Fig. 2. It is evident that the proposed peak detection

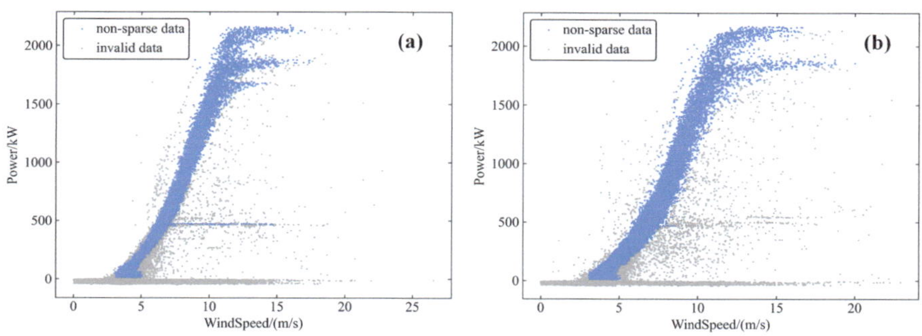

Fig. 1 The sparse invalid data identification results. **a** Wind turbine 1, **b** Wind turbine 8

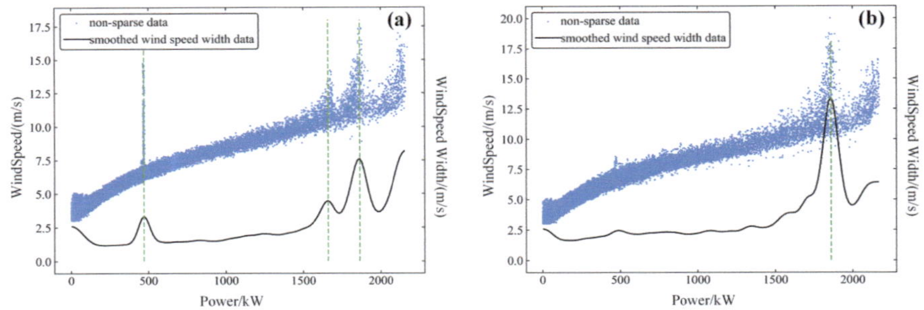

Fig. 2 Verification of data smoothing feasibility. **a** Wind turbine 1, **b** Wind turbine 8

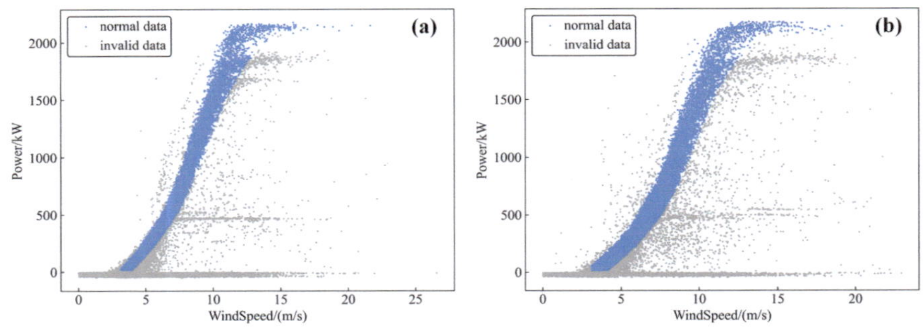

Fig. 3 The results of invalid data identification. **a** Wind turbine 1, **b** Wind turbine 8

algorithm accurately captures the locations of the peaks formed by downrating invalid data, along with the trend of wind speed width data as it varies with power.

The invalid wind power data identification results are presented in Fig. 3a, b. As evident from Fig. 3, the proposed method exhibits excellent recognition performance for various types of invalid data.

3.2 Identification and Verification of Artificial Simulation Data

To quantitatively analyze the recognition performance of the proposed method in comparison to existing methods, this paper simulates artificial invalid data based on the filtered normal data from turbine 1. To validate the effectiveness of the proposed method under varying proportions of three types of invalid data, the recognition rate R_m, the overall invalid data recognition rate R_s, and the false recognition rate F are defined as follows:

$$R_m = \frac{n_{mr}}{n_m} \tag{7}$$

$$R_s = \frac{n_{fr} + n_{dr} + n_{lr}}{n_f + n_d + n_l} \tag{8}$$

$$F = \frac{n_{sf}}{n_s} \tag{9}$$

where $m = f, d, l$, represent invalid data of bottom stack type, sparse type, and downrating type, respectively, n_m denotes the total number of invalid data of type m, and n_{mr} represents the number of invalid data of type m identified by the invalid data recognition method. n_s represents the total number of normal data, and n_{sf} represents the number of normal data that are misidentified as invalid data.

a) **Recognition Performance Under Different Proportions of Invalid Data**

The recognition performance of the proposed method under various ratios of the three types of invalid data is shown in Table 1.

Based on Table 1, the recognition rates of the proposed method for various types of invalid data are all above 95%. This demonstrates the excellent recognition performance of the method for the three categories of invalid data. Furthermore, regardless of the different proportions of the three types of invalid data, the proposed method maintains good recognition accuracy, validating its generality and applicability.

b) **Comparative Analysis with Other Methods**

The widely used quartiles, DBSCAN, and LOF methods are compared with the proposed method in this paper. The simulated data is visualized in Fig. 4a. For the DBSCAN-based method, the parameters are selected using the adaptive DBSCAN algorithm proposed in [14], with Eps set to 0.01 and Minpts set to 30. The parameter *n_neighbors* for LOF is set to 3% of the total data volume, as suggested in [15]. The recognition results for each method are presented in Fig. 4 and Table 2.

Figures 4b–e depict the recognition results of invalid data using the proposed method, the quartile-based method, DBSCAN-based method, and LOF-based method, respectively. As observed in Fig. 4 and Table 2, it is evident that the proposed invalid data identification method outperforms the other three methods. The quartile method is better suited for data with symmetric distribution but demonstrates poorer recognition performance for downrating invalid data. When the density of normal data closely resembles that of invalid data, the DBSCAN method may incorrectly classify a significant portion of normal data as invalid. Conversely, the LOF method

Table 1 Identification results of invalid data with different proportions

n_s	n_f	n_d	n_l	$R_f/\%$	$R_d/\%$	$R_l/\%$	$R_s/\%$
26,667	2000	200	525	100	99.5	98.86	99.97
26,667	2000	1200	357	100	99.58	96.36	99.49
26,667	2000	3000	728	100	96.3	95.33	97.47

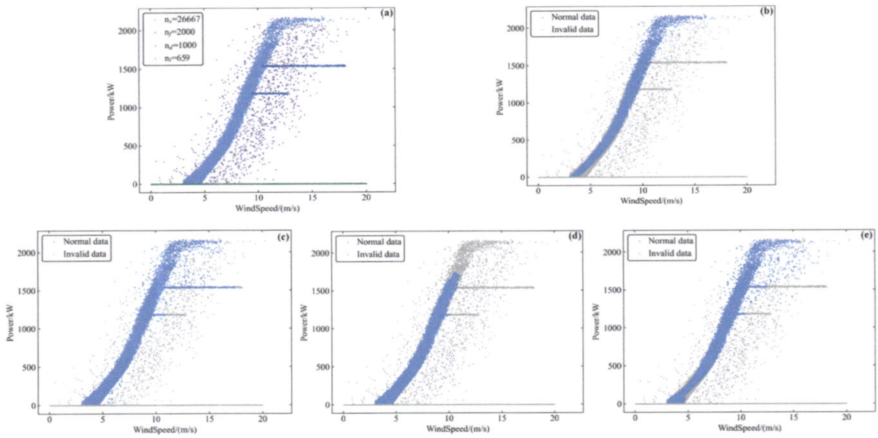

Fig. 4 The artificial simulation data and invalid data identification results of different methods on simulated data. **a** The artificial simulation data, **b** proposed method, **c** the quartile method, **d** DBSCAN, **e** LOF

Table 2 Identification results comparison of different methods

	R_f/%	R_d/%	R_l/%	R_s/%	F/%
Proposed	100	96.3	96.63	98.40	1.68
Quartile	100	85.6	77.39	91.99	0.06
DBSCAN	100	98	54.32	91.23	3.02
LOF	100	68.8	68.29	85.76	1.49

identifies outliers by comparing the local density of data points with their neighbors, which is evidently impractical for detecting downrating invalid data.

4 Conclusion

We propose a method to identify invalid wind power data using segmented quartiles and peak detection in this paper. Parameters related to the quartile method can be automatically calculated based on the input data, eliminating the need to manually adjust parameters for different datasets. Based on the morphological characteristics of the invalid data of downrating, the peak detection algorithm is used to identify the downrating invalid data, which solves the weakness of the traditional method that is difficult to identify the downrating invalid data. Compared with the traditional method, the proposed algorithm has a better invalid data recognition effect. Future work will focus on developing detection methods for scenarios with a high proportion of sparse invalid data and optimizing the fitting effect of datasets.

References

1. Wang, Y., Zhu, N., Xie, H., et al.: Ultra-short-term wind power prediction method based on contrastive learning assisted training. Chin. J. Sci. Instrument **3**, 89–97 (2023)
2. Wang, Y., Hu, Q., Li, L., et al.: Approaches to wind power curve modeling: a review and discussion. Renew. Sustain. Energy Rev. **116**, 109422 (2019)
3. Yan, J., Zhang, H.: Uncertainty estimation for wind energy conversion by probabilistic wind turbine power curve modelling. Appl. Energy **239**, 1356–1370 (2019)
4. Sun, C., Guo, P.: Data preprocessing of wind turbine based on least squares support vector machine and neighbor model. In: 29th Chinese Control and Decision Conference (CCDC), pp. 1441–1446. IEEE, Chongqing, China (2017)
5. Zou, T., Gao, Y., Yi, H., et al.: Processing of wind power abnormal data based on Thompson tau-quartile and multi-point interpolation. Autom. Electric Power Syst. **44**(15), 156–162 (2020)
6. Long, H., Sang, L.: Image-based abnormal data detection and cleaning algorithm via wind power curve. IEEE Trans. Sustain. Energy **11**(2), 938–946 (2020)
7. Liang, G., Su, Y., Chen, F., et al.: Wind power curve data cleaning by image thresholding based on class uncertainty and shape dissimilarity. IEEE Trans. Sustain. Energy **12**(2), 1383–1393 (2021)
8. Li, Y., Shen, X., Zhang, Y., et al.: Cleaning method of wind speed outliers for wind turbines based on velocity and correlation constraints. Trans. China Electrotechnical Soc. **38**(7), 1793–1807 (2023)
9. Qin, Y., Mei, J., Ren, H., et al.: Image reconstruction based on Gaussian smoothing compressed sensing fractional order total variational algorithm. J. Electron. Inf. Technol. **43**(7), 2105–2112 (2021)
10. Bi, J., Wang, Y., Cao, H., et al.: A smart phone step counting algorithm for peak and trough detection. J. Chin. Inertial Technol. **28**(3), 287–292 (2020)
11. Feng, G., Lai, C., Han, Y., et al.: Fast Maximum Torque Per Ampere (MTPA) angle detection for interior PMSMs using online polynomial curve fitting. IEEE Trans. Power Electron. **37**(2), 2045–2056 (2022)
12. Zhou, J., Wang, Y., Hisakazu, K.: Doppler power spectrum density and multi-antenna system performance in three-dimensional environment. Acta Physica Sinica **63**(24), 240507–240507 (2014)
13. Wu, X., Sun, R., Qiao, Y., et al.: Estimation of error distribution for wind power prediction based on power curves of wind farms. Power Syst. Technol. **41**(6), 1801–1807 (2017)
14. Hu, J., Lai, J., Li, Y., et. al.: Evaluation method of switchgear insulation state based on adaptive density based spatial clustering of applications with noise algorithm. Trans. China Electrotechnical Soc. **36**(zk1), 344–352 (2021)
15. Liu, S., Zhao, Y., Lin, Z., et al.: Data-driven event detection of power systems based on unequal-interval reduction of PMU data and local outlier factor. IEEE Trans. Smart Grid **11**(2), 1630–1643 (2020)

An Improved Retinex-Net Low-Light Image Enhancement Method

Zhiqiang Zhao, Yuanfei Wang, Xinyuan Ouyang, and Yaofang Lu

Abstract Noise and false exposure are important reasons for image information loss. Aiming at the problem that it is difficult to control exposure and noise in low-light image enhancement, a multi-branch convolutional network is designed and the loss function is modified to enhance image details and remove noise. Inspired by the MSR algorithm, the algorithm designs a three-branch five-layer convolution kernel network structure, and avoids underexposure or overexposure that may occur in a single branch by means of multi-branch averaging. In order to balance noise removal and detail preservation, the SSIM loss function is introduced, combined with the U-Net like network structure of the residual network, to compress image features to reduce the proportion of noise while retaining complete structural information. Experimental results verify the effectiveness of the proposed network.

Keywords Low light images · Image enhancement · Retinex-Net · Denoising

1 Introduction

In real-life scenarios, low light, backlighting, and uneven illumination often occur, resulting in images with low brightness, low contrast, and a narrow grey scale range. These images may also be accompanied by noise, distortion, and other phenomena that can greatly affect the visual quality, accurate conveyance of information, and downstream tasks. Therefore, it is necessary to enhance low-light images.

According to the Retinex theory [1], the image perceived by the human eye comprises a reflected image that represents the object's nature and a lighted image. The theory suggests that the object's colour is not related to the intensity of the light but is determined by its ability to reflect waves of three different lengths. Furthermore,

Z. Zhao (✉) · X. Ouyang · Y. Lu
School of Software Engineering, Chongqing University of Posts and Telecommunications, Chongqing, China
e-mail: zhaozq@cqupt.edu.cn; S220332006@stu.cqupt.edu.cn

Y. Wang · X. Ouyang · Y. Lu
Chongqing University of Posts and Telecommunications, Chongqing, China

J. S. Park et al. (eds.), *Advanced Multimedia and Ubiquitous Engineering*, Lecture Notes in Electrical Engineering 1475, https://doi.org/10.1007/978-981-95-1565-3_17

the object's colour remains constant even when the light changes. Wei [2] created the first low-light image dataset (LOL) for real scenes and designed a deep convolutional neural network called Retinex-Net based on Retinex theory. The network comprises a decomposition network and an illumination map enhancement network. A loss function is designed to guide the network learning to obtain the mapping function between the low-illumination image and the reference image. The resulting enhanced image shows significant improvements in brightness, contrast, and detail compared to previous methods. Jiang [3] applied Generative Adversarial Nets (GAN) to enhance low illumination images. Through a constant game between generator and discriminator, high-quality images with good visual effects were generated. Guo [4] designed several loss functions, including exposure control, color constancy, illuminance smoothing, and spatial consistency. The network is designed to train the model using these loss functions, without relying on image pairs. The resulting model can adapt to multiple scenes, and its generalization ability has been greatly improved.

This paper proposes ideas based on Retinex theory and Retinex-Net. As images with diverse resolutions can cause overexposure, the Retinex-Net network is unable to cope with this phenomenon. To address this issue, a multi-branch network is used to fuse information and perform illumination map enhancement. The reflection network utilises a combination of the U-Net [5] and residual network to compress features. The loss function is improved to balance detail preservation and noise removal by introducing SSIM. This results in a denoised reflection map, which is combined with the enhanced illumination map to produce the final enhanced image.

2 Retinex-Net and Its Issues

Retinex-Net comprises of a decomposition network (Decom) for decomposition and an end-to-end enhancement network for illumination adjustment (Relight). The Decom network is trained without decomposing the reference maps of the reflectance and illumination maps. The network is learned with only key constraints, including consistent reflectance shared by low/normal light images, and smoothness of illumination. The light is enhanced for luminance using an enhancement network, and the reflectance is then noise-reduced by joint denoising, based on the decomposition.

Retinex-Net utilises a CNN-based decomposition and enhancement process. During Decom-Net training, constraints are introduced to ensure consistency in the reflection map and smoothness in the illumination map. However, due to limitations in Decom-Net, complete consistency cannot be achieved for the reflection components in low light/normal light images, resulting in blurred images. However, the illumination map produced by Relight may result in over-enhancement, causing certain areas of the final image to appear excessively bright and white. The BM3D [6] algorithm is employed to remove noise from the reflection map. It combines spatial and transform domain methods to achieve better noise removal. However, the algorithm is computationally intensive and complex.

3 Retinex-Net Model Improvement

3.1 Illumination Map Enhanced Network

Retinex theory proposes that a low-light image comprises of a reflection map and an illuminance map. The reflectance map represents the inherent properties of the image, independent of the lighting conditions, while the illuminance map determines the dynamic range of the pixels in the image, so any under or over-adjustment of the illuminance map will affect the enhancement of the visual effect of the image. This paper aims to improve the Retinex-Net illuminance map enhancement network by addressing its shortcomings. The proposed approach is inspired by the MSR [7] algorithm and involves designing a multi-branch network with different sizes of convolutional kernels to adjust the illuminance map (Fig. 1).

The network structure has been improved by utilising multiple branches with convolution kernels of varying sizes. Two considerations were taken into account. Firstly, when the convolution kernel is very small, it provides good edge enhancement but poor results in the gentle zone. Conversely, when the kernel is large, it provides poor edge enhancement but good results in the gentle zone. To address this, multi-scale convolution kernels were used to capture different sensory fields and improve global feature capture. Secondly, by summing up the multi-branch features for the mean value, the over-adjustment phenomenon that may occur in a single branch can be avoided, preventing over-enhancement of the results. The processing can be expressed as follows:

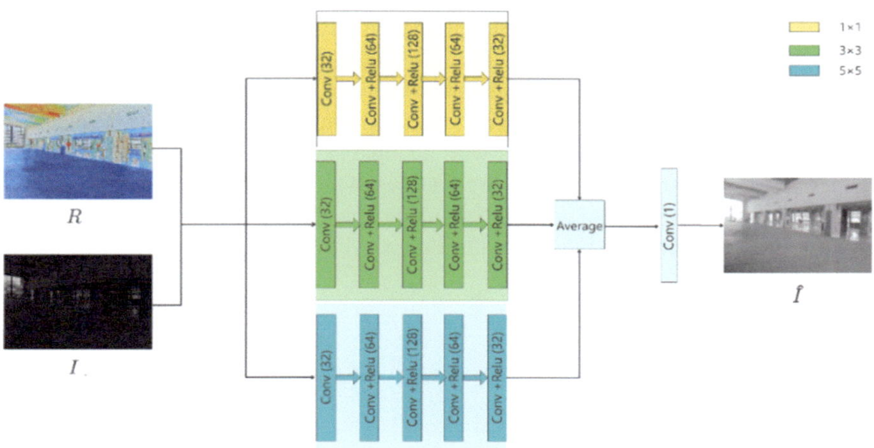

Fig. 1 Improved illumination map enhancement network. The reflection and illuminance maps, denoted as R and I respectively, were obtained from the input decomposition. The adjusted illuminance map, $\hat{I}$, was also obtained

$$r(x, y) = \sum_{j=1}^{3} w_j \{\log I(x, y) - \log[F(x, y) * I(x, y)]\} \quad (1)$$

where $r(x, y)$ is the logarithmic representation of the reflection map, $I(x, y)$ is the original image, $F(x, y)$ is the Gaussian kernel function, and * represent the convolution operation. The network for enhancing the illumination map has three branches, each using a different convolutional kernel. The features obtained from each branch are then summed up element by element for averaging. Finally, a convolutional layer is applied to obtain the adjusted illumination map.

3.2 Reflection Map Denoising Network

This paper presents a network designed to balance image detail enhancement and denoising, with a focus on the latter. To avoid loss of details caused by denoising, the paper introduces SSIM as a loss function. The network takes the reflection map obtained by decomposition as input and outputs the reflection map after detail enhancement and denoising. The denoising network is modelled as (Fig. 2).

To enhance network training speed and improve denoising in the reflection map, the denoising network utilises a residual block plus U-Net structure.

The process of network processing for enhancing images can be summarised as follows: Firstly, the network is fed with the reflection map obtained from decomposition. The network training is then constrained by the loss function, and the network weights are adjusted using the backward propagation algorithm to obtain detail-enhanced and denoised reflection maps. Finally, the adjusted illuminance map and the denoised reflection map are multiplied together to obtain the final enhanced image. The residual block can prevent the problem of gradient disappearance caused by deeper networks, allowing for faster convergence to the optimal value. The convolution block can be seen as a U-Net with a single skip connection. This is because deeper networks extract more advanced features, which refer to the detailed features in the image. Therefore, the convolution block can fuse rich detailed features. The up-sampling layer can double the feature map, while the splicing layer can merge

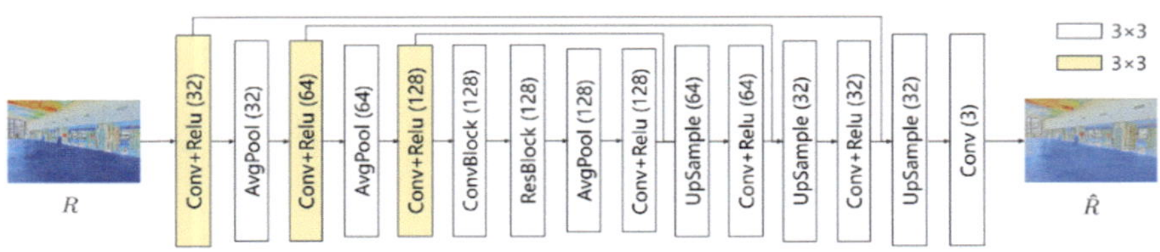

Fig. 2 Reflection map denoising network model. The reflectogram obtained from decomposition is represented by R, while the optimised reflectogram is represented by $\hat{R}$

the shallow and high-level features of the same level. Finally, a convolution layer can be used to obtain the feature map with the desired number of channels.

The loss function for the reflection map noise reduction network comprises three components: reflection map consistency loss, reconstructed image loss, and structural similarity loss. The total loss function is defined in Eq. (2).

$$L_{remove\_noise} = \lambda_{tr} L_{ir} + \lambda_{recon} L_{recon} + \lambda_{ssim} L_{ssim} \tag{2}$$

λ_{ir} has a value of 0.01 and λ_{recon} has a value of 1. After conducting multiple tests, λ_{ssim} was found to be 5. Equation (3) shows the loss of reflectogram consistency.

$$L_{ir} = \left\| R_{normal} - \hat{R}_{low} \right\|_1 \tag{3}$$

R_{normal} and $\hat{R}_{low}$ represent the reflection maps of the reference image and the noise-reduced image, respectively. Equation (4) represents the image reconstruction loss.

$$L_{recon} = \left\| \hat{R}_{low} \circ \hat{I}_{low} - S_{normal} \right\|_1 \tag{4}$$

The adjusted illumination map is represented by $\hat{I}_{low}$, and pixel-by-pixel multiplication is denoted by '∘'. Equation (5) represents the loss of structural similarity.

$$L_{ssim} = SSIM\left(R_{high}, \hat{R}_{low}\right) \tag{5}$$

The SSIM loss in this case is calculated using Wang's method [8].

4 Experimental Results

To verify the effectiveness of the algorithm in this paper, two low-light image test sets are selected for experimental comparison: the LOL [2] paired test set and the LIME [9] unpaired test set. Four representative low light imaging algorithms from recent years were selected for comparison, including Retinex-Net [2], EnlightenGAN [3], Zero-DCE [4], and RetinexDIP [10]. Retinex-Net and RetinexDIP are research algorithms based on Retinex. EnlightenGAN is an unsupervised low-light image algorithm that uses a GAN network. Zero-DCE is a zero-reference image enhancement algorithm with strong generalisation ability. The algorithms were designed using the recommended settings of the paper for all parameters.

Fig. 3 Comparison of low-illumination image processing algorithms. **a** Original. **b** Retinex-Net [2]. **c** EnlightenGAN [3]. **d** Zero-DCE [4]. **e** Ours

4.1 Subjective Comparison

To assess the effectiveness of the algorithm presented in this paper, the classic LOL test set and the commonly used real low-light image dataset are selected for low light image enhancement.

In Fig. 3, distortions in the images produced by Retinex-Net, including the leaf in the lower right corner of the second set of images in (b). The results of EnlightenGAN also exhibit distortions in the middle part. Zero-DCE produces images that are generally darker in appearance. Ours algorithm outperforms the un-noise reduction algorithm in subjective vision. The enhanced images appear more natural and align better with human visual perception.

4.2 Objective Comparison

To ensure objective evaluation of the algorithm results, five objective evaluation metrics are selected: Peak Signal to Noise Ratio (PSNR) [11], Structural Similarity (SSIM) [12], Lightness order error (LOE) [13], Perceptual Index (PI) [14], and Natural Image Quality Evaluator (NIQE) [15].

Table 1 shows the evaluation metrics averaged over the LOL dataset, with the best performing values highlighted in bold black font. The PSNR metrics indicate that this paper's algorithm has the highest value for noise removal. Additionally, the algorithm has the highest SSIM value, indicating that the enhanced image performs better in terms of brightness, contrast, and structure, and is more similar to the reference image. The algorithm described in this paper achieved the second-best result on the no-reference evaluation metric LOE, NIQE. This result indicates that the enhanced image has better perceptual quality and is consistent with human visual perception.

The evaluation metrics in Table 2 are averaged over the LIME dataset. In the LOE evaluation index, ours rank second, which is slightly inferior to Zero-DCE and much higher than Retinex-Net. In the PI index, ours also rank second, indicating that the enhanced image obtained by the integrated improved model has a better visual effect.

Table 1 Comparison of objective evaluation indicators of LOL data set

Model	PSNR↑	SSIM↑	LOE↓	PI↓	NIQE↓
Retinex-Net	19.3847	0.5816	486.2084	8.6714	8.8789
EnlightenGAN	19.6945	0.7638	366.1990	8.8918	**4.6839**
Zero-DCE	16.6889	0.7335	**216.5900**	8.8977	7.7925
RetinexDIP	11.0798	0.5399	516.3767	**6.7024**	6.9807
Ours	**20.5342**	**0.8405**	304.1320	8.8337	6.8795

Table 2 Comparison of objective evaluation indicators of LIME data set

Model	LOE↓	PI↓	NIQE↓
Retinex-Net	589.6400	8.3802	4.5977
EnlightenGAN	421.1000	**7.8866**	**3.6574**
Zero-DCE	**187.2200**	8.1762	3.8022
RetinexDIP	305.0300	8.2035	3.8088
Ours	275.6585	8.1691	4.1973

The above comparison shows that the integrated model not only has a better effect in denoising but also performs well in terms of visual effect.

5 Conclusion

This paper aims to address issues with the Retinex-Net low-light image enhancement algorithm by studying low-light images from two perspectives. The first perspective involves improving brightness and contrast. To achieve this, the paper draws inspiration from the MSR algorithm and builds a three-branch five-layer network structure on top of the Retinex-Net network model. This approach results in enhanced images that perform better both subjectively and objectively. The second aspect is to balance the details of the image and remove noise. This can be achieved by designing a reflection map noise reduction network with residual blocks and a U-Net-like structure. This will allow the network to quickly converge to the optimal value while retaining the image details, resulting in improved detail enhancement and noise removal. The effectiveness of this improvement was verified through model comparison experiments.

References

1. E Land 1997 The retinex theory of color vision Sci. Am. 237 6 108 128
2. Wei, C., Wang, W.J., Yang W.H.: Deep retinex decomposition for low-light enhancement. In: British Machine Vision Conference (BMVC). ACM, Newcastle, UK (2018)
3. YF Jiang XY Gong D Liu 2021 EnlightenGAN: deep light enhancement without paried supervision IEEE Trans. Image Process. 30 2340 2349
4. Guo, C.L., Li, C.Y., Guo, J.C.: Zero-reference deep curve estimation for low-light image enhancement. In: Conference on Computer Vision and Pattern Recognition (CVPR), pp 1777–1786. IEEE, Seattle, WA, USA (2020)
5. Ronneberger, O., Fischer, P., Brox, T.: U-Net: convolutional networks for biomedical image segmentation. In: Medical Image Computing and Computer-Assisted Intervention, pp. 234–241. ACM, Munich, Germany (2015)
6. M Lebrun 2012 An analysis and implementation of the BM3D image denoising method Image Process. On Line 2 175 213
7. D Jobson Z Rahman G Woodell 1997 A multiscale retinex for bridging the gap between color images and the human observation of scenes IEEE Trans. Image Process. 6 7 965 976
8. Z Wang A Bovik H Sheikh 2004 Image quality assessment: from error visibility to structural similarity IEEE Trans. Image Process. 13 4 600 612
9. XJ Guo L Yu HB Ling 2017 LIME: low-light image enhancement via illumination map estimation IEEE Trans. Image Process. 26 2 982 993
10. ZJ Zhao BS Xiong L Wang 2022 RetinexDIP: a unified deep framework for low-light image enhancement IEEE Trans. Circuits Syst. Video Technol. 32 3 1076 1088
11. Galdran, A., Bria, A., Alvarez, G.: On the duality between retinex and image dehazing. In: Conference on Computer Vision and Pattern Recognition, pp. 8212–8221. Salt Lake City, UT, USA (2018)
12. Zhou, W., Bovik, A.C., Sheikh H.R.: Image quality assessment: from error visibility to structural similarity. IEEE Trans. Image Process. 600–612 (2004)
13. Wang, S., Zheng, J., Hu, H.M.: Naturalness preserved enhancement algorithm for non-uniform illumination images. IEEE Trans. Image Process. 3538–3548 (2013)
14. C Ma CH Yang 2017 Learning a no reference quality metric for single-image super-resolution Comput. Vis. Image Underst. 158 5 1 16
15. Mittal, A., Soundararajan, R., Bovik, A.: Making a "Completely Blind" image quality analyzer. In IEEE Signal Processing Letters, pp. 209–212 (2013)

User Portrait-Based Cabin Thermal Comfort Temperature Decision Algorithm for New Energy Vehicles

Yongwei Li, Yunpeng Shi, Wenqiang Li, and Yang Luo

Abstract In order to further improve the intelligence and comfort level of the thermal system of automobile cabins, a user portrait-based cabin thermal comfort temperature decision algorithm is proposed. The age, gender, body mass index (BMI), cloth class and height features of users are obtained by designing three neural networks. Subsequently, the predicted percentage of dissatisfaction (PPD) index is calculated according to the obtained 5 features. With using the iteration optimization strategy, the cabin thermal comfort temperature is generated with satisfying the ISO7730 standard. Experimental test results show that the proposed method can efficiently reduce the PDD index so as to satisfy the thermal comfort, revealing the application value of the user portrait-based decision strategy.

Keywords Automobile · Thermal comfort · User portrait · Temperature decision

1 Introduction

Thermal comfort for new energy vehicles is very important [1]. Currently, the adjustment of the thermal conditioning in the car cabins mainly relies on the judgment of drivers or the historical memory. Meanwhile, the thermal conditioning system monitors the indoor temperature through built-in sensors and uses fuzzy control or PID

Y. Li (✉) · Y. Shi · W. Li
Chongqing Changan Automobile Co.,Ltd, Chongqing, China
e-mail: liyw3@changan.com.cn

Y. Shi
e-mail: shiyp@changan.com.cn

W. Li
e-mail: liwj5@changan.com.cn

Y. Luo
College of Automation, Chongqing University of Posts and Telecommunications, Chongqing, China
e-mail: s210332015@stu.cqupt.edu.cn

J. S. Park et al. (eds.), *Advanced Multimedia and Ubiquitous Engineering*, Lecture Notes in Electrical Engineering 1475, https://doi.org/10.1007/978-981-95-1565-3_18

control algorithms [2] to automatically adjust the working status of the air conditioning system so as to approach the set target temperature. Furthermore, in the field of automotive thermal conditioning research, the researchers mainly focus on energy consumption optimization, improvement of refrigeration and heating capacity, etc.

In the past decades, the thermal research trends of automobiles include intelligence, networking, and personalization. New ideas and methods have also been proposed for personalized thermal comfort intelligent control of car cabins. For instance, Khatoon et al. [3] analyzed the cabin thermal comfort by using the three-demission model and the temperature and velocity of the cabin were accurately predicted. Alok et al. [4] employed the data-driven machine learning method to construct the car cabin thermal comfort model and achieved high precision real-time decision making. However, by analyzing current researches, it is also found that the satisfaction level of personalized thermal comfort for drivers and passengers is still relatively low. To further improve the thermal comfort performance of car cabin, a user portrait-based cabin thermal comfort temperature decision algorithm is proposed in this work. The feature extraction networks are employed to obtain 5 important features. Subsequently, with using the obtained features and introducing the predicted percentage of dissatisfaction (PPD) index, the cabin thermal comfort temperature calculation method is designed. Experimental tests are conducted to verify the performance of the proposed method.

2 User Portrait Feature Extraction

2.1 YOLOv5-Based Feature Extraction

As one of the most efficient object detection methods, YOLOv5 has been widely used. To obtain the face coordinate box, cloth and height features of users, the YOLOv5-based feature extraction is introduced and the frame is given in Fig. 1. It can be seen in Fig. 1 that there are 4 sections in YOLOv5 network, including input module, backbone module, neck module and prediction output module. To grantee high object detection precession, the adaptive training sample selection algorithm is used in the input module. Subsequently, the focus structure is employed in the backbone module to execute convolution operation so as to reduce parameters and computation cost. Accordingly, the neck module uses the path aggregation network to improve the perception ability for different scales targets. Finally, the prediction information can be obtained in the output module, where the confidence score, class probability and boundary box coordinates are provided.

Based on the output prediction information, the face coordinate box and cloth class then can be obtained. However, since the passengers always sit on the seat in the cabin, it is difficult to calculate the accurate height feature. To tackle this issue, the pinhole model [5] is employed to assume the height information by the formula below:

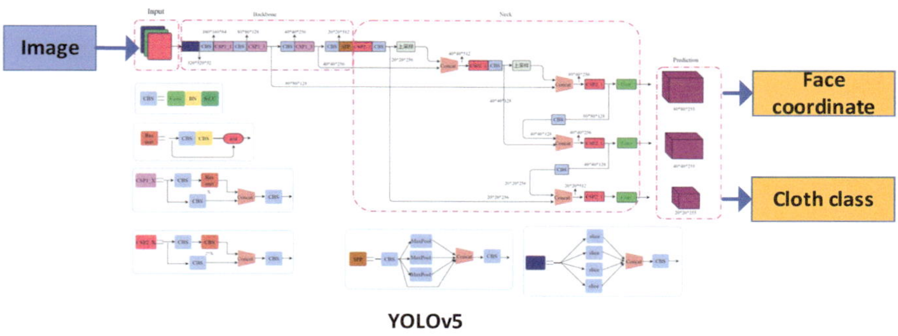

Fig. 1 Frame of the YOLOv5-based feature extraction

$$H_{ei} = 7.5 \times F_{head} \times d/f \quad (1)$$

where F_{head} means the length of the face coordinate box, d is the distance between the passenger and camera, f is the focal distance of the camera.

2.2 *Multi-task Learning-Based Feature Extraction*

Multi-task learning (MTL) is a kind of machine learning method, which learns multiple missions during one model. By using the face coordinate box information as the input, the MTL module is employed in this work to further obtain the age and gender features. Briefly, the multi-mission learning model is employed and the structure of this model is shown in Fig. 2. In this model, the cross-entropy loss is used as the loss function and the sharing part of the network is ResNet-50. In the non-shared part, two layers are contained and fully connection layers are followed. Accordingly, the age and gender features are obtained by using the employed MTL network (Table 1).

2.3 *Openpose-Based Feature Extraction*

Since the BMI feature cannot be obtained directly, the human posture extraction is employed to estimate this parameter. Therefore, the Openpose network developed from convolutional neural network (CNN) and supervised learning (SL) is improved in this work, where the ResNet50 is used as the backbone network to construct the human body key points detection model. Meanwhile, during the process of feature extraction, the middle point of the human body key points and the distance between these points are calculated by the following two formulas:

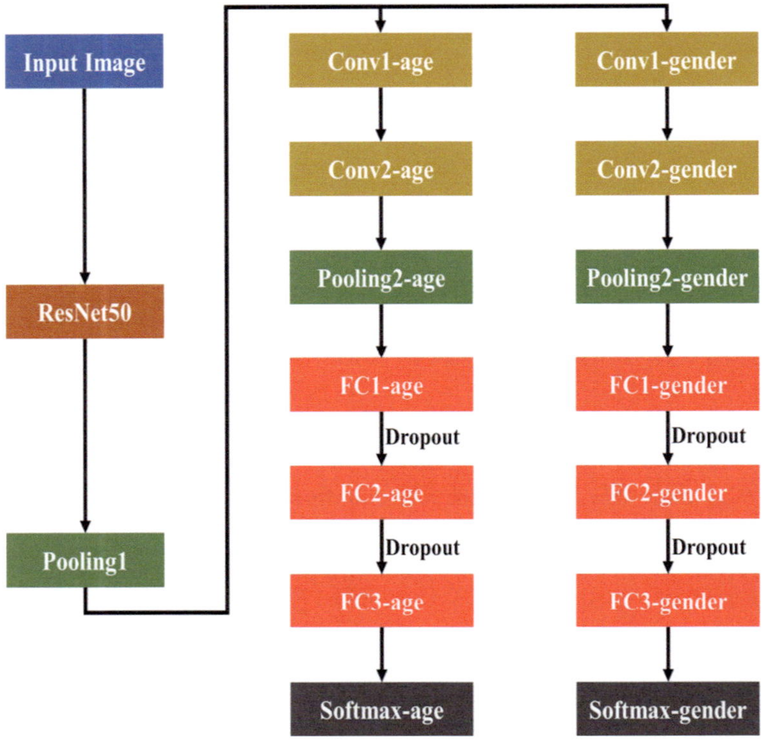

Fig. 2 Structure of the multi-mission learning MTL model

Table 1 Parameters of the multi-mission learning MTL model

Layer name	Layer type	Dimension	Layer name	Layer type	Dimension
ResNet50	ResNet	2048 * 7 * 7	FC1-age/gender	Connection	1 * 1024
Pooling1	Avg-pooling	2048 * 7 * 7	FC2-age/gender	Connection	1 * 512
Conv1-age/ gender	Convolution	1024 * 7 * 7	FC3-age	Connection	1 * 7
Conv2-age/ gender	Convolution	1024 * 5 * 5	FC3-gender	Connection	1 * 2
Pooling2-age/ gender	Avg-pooling	1024 * 1 * 1			

$$Middle\,(p1, p2) = \left[(p1.x + p2.x)/2, (p1.y + p2.y)/2\right] \tag{2}$$

$$Length\,(p1, p2) = \sqrt{(p1.x - p2.x)^2 + (p1.y - p2.y)^2} \tag{3}$$

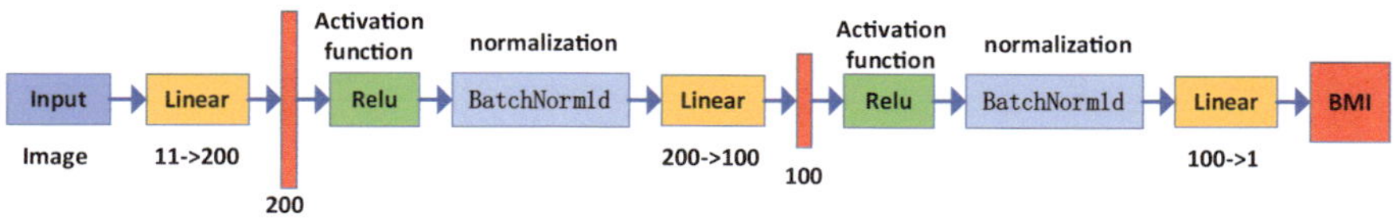

Fig. 3 Structure of the BMI detection neural network model

where $p1$ and $p2$ mean the two body key points, $(p1.x, p1.y)$ and $(p2.x, p2.y)$ mean the coordinates of the two points. On this basis, the BMI detection neural network model is proposed in this work and the structure is shown in Fig. 3. Finally, the BMI feature is obtained by using the proposed model.

3 Cabin Thermal Comfort Temperature Computation Method

With using the feature extraction, the corresponding 5 features can be obtained and are defined in Table 2. However, each passenger has their own thermal comfort sense. To calculate the cabin thermal comfort temperature, an improved user portrait-based temperature computation method is proposed in this work.

Firstly, the energy metabolism rate of human is calculated below:

$$M = \begin{cases} \dfrac{0.08 \times (9.56 \times W_{BMI} + 1.85 \times H_{ei} - Y \times 4.67 + 655.09)}{0.0073 H_{ei} + 0.0121 W_{BMI} - 0.2106}, \text{Female} \\ \dfrac{0.08 \times (13.75 \times W_{BMI} + 5.0 \times H_{ei} - Y \times 6.76 + 664.73)}{0.0057 H_{ei} + 0.0121 W_{BMI} + 0.0882}, \text{Male} \end{cases} \tag{4}$$

where M is the energy metabolism rate of human (W/m^2). Subsequently, the human heat load in cabin is calculated according to the following formula:

$$\begin{aligned} L = M &- 3.054 \times 10^{-3} \times \left[5733 - 6.99M - RH \times e^{\left(16.6 - \frac{4030.1}{t+235}\right)}\right] - 0.42 \\ &\times (M - 58.15) - 1.7 \times 10^{-5} M \times \left(5867 - RH \times e^{\left(16.6 - \frac{4030.1}{t+235}\right)}\right) \\ &- 0.0014 M (34 - t) - 3.96 \times S \end{aligned}$$

Table 2 Passenger feature definition

Feature name	Simbel	Network	Feature name	Simbel	Network
Cloth class	$\{S_{clo}, SA_{clo}, W_{clo}\}$	YOLOv5	Height	H_{ei}	YOLOv5
Gender	{female, male}	MTL	BMI	W_{BMI}	BMI detection
Age	Y	MTL			

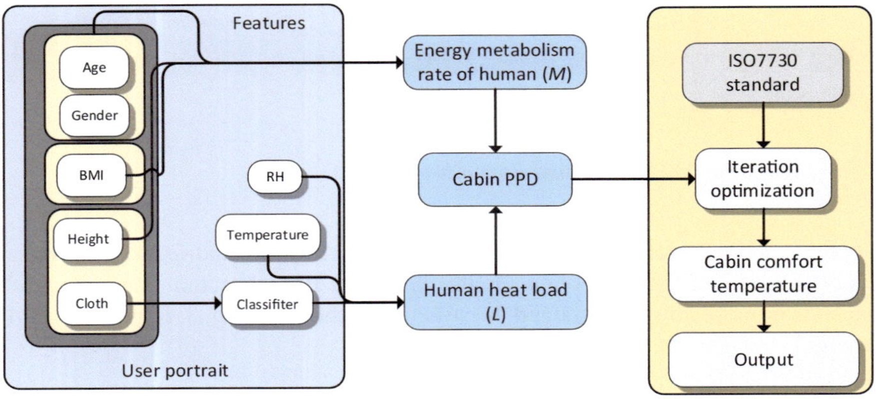

Fig. 4 Flow chart of the cabin thermal comfort temperature computation method

$$\times \left[((35.7 - 0.028M + 273)/100)^4 - ((t + 273)/100)^4\right] \tag{5}$$

where RH is the humidity of the cabin (%), t is the current temperature of the cabin and $S \in \{S_{clo}, SA_{clo}, W_{clo}\}$. Next, the PPD index [6] is introduced to judge the cabin thermal comfort performance:

$$C_{PPD} = 100 - 95 \times e^{-(0.03\times PMV^4 + 0.22\times PMV^2)} \tag{6}$$

where C_{PPD} is the cabin PPD index. Based on the ISO7730 standard [7], it is expected that the PPD index is less than 10% so that the temperature is comfortable for humans. Therefore, the cabin thermal comfort temperature can be calculated by the using the Eq. (6). Briefly, the flow chart of the cabin thermal comfort temperature computation method is given in Fig. 4.

4 Experimental Tests

To verify the performance of the proposed computation method, an experimental platform is constructed and the frame diagram is shown in Fig. 5. In this platform, the camera in the new energy vehicle is employed to obtain the passenger images and then transform these images into the TBOX computation platform, where the proposed cabin thermal comfort temperature computation method is deployed. Meanwhile, the temperature/humidity sensor is used to detect the temperature and humidity features. When the computation platform obtains the image and physical parameters, the proposed method is executed to generate the thermal comfort temperature, and the results are sent to the cabin control platform.

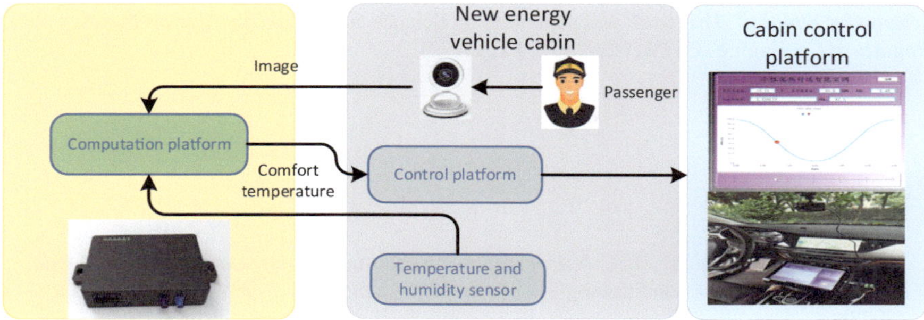

Fig. 5 Structure of the experimental platform

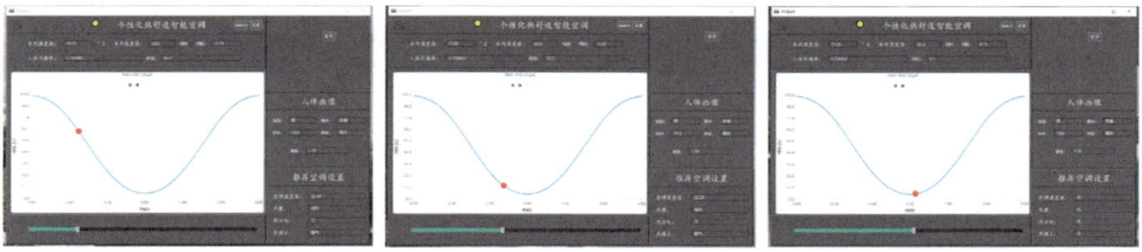

(a) Initial status (b) Adjustment status (c) Final status

Fig. 6 Experiential test results

The experiential test results are shown in Fig. 6. It can be seen that when the camera and sensor transfer the passenger image and temperature and humidity parameters to the computation platform, the initial cabin PPD index is 64.1%. By using the proposed method, a thermal comfort temperature of 22 °C is recommended. During the adjustment, it can be seen that the PPD index decreases from 64.1 to 12.7.1%. Finally, the PPD index stabilizes at 5.5%. Test results shows that the proposed method can efficiently calculate the thermal comfort temperature, revealing the application value for new energy vehicles.

5 Conclusion

In this work, a user portrait-based cabin thermal comfort temperature decision algorithm is proposed. Three neural networks are designed to obtain user features. Accordingly, the PPD index can be calculated and then the cabin thermal comfort temperature is obtained with using iteration optimization strategy. Experimental test results demonstrate that the proposed method can generate thermal comfort temperature, which satisfies the ISO7730 standard. The extension of the proposed method can be intelligent temperature decision and control.

Acknowledgements This work was supported by Science and Technology Innovation Key R&D Program of Chongqing (CSTB2023NSCQ-LZX0172)

References

1. Huang, X., Li, K., Xie, Y., et al.: A novel multistage constant compressor speed control strategy of electric vehicle air conditioning system based on genetic algorithm. Energy **241**(2), 1–16 (2022)
2. Li, X., Chen, J., Zhou, D., et al.: A modified biogeography-based optimization algorithm based on cloud theory for optimizing a fuzzy PID controller. Optimal Control Appl. Methods **3**, 43 (2022)
3. Khatoon, S., Kim, M.H.: Thermal comfort in the passenger compartment using a 3-D numerical analysis and comparison with Fanger's comfort models. Energies **13**(3), 1–15 (2020)
4. Warey, A., Khalighi, B.: Data-driven prediction of vehicle cabin thermal comfort: using machine learning and high-fidelity simulation results. Int. J. Heat Mass Transf. **148**, 119083 (2019)
5. Ren, S., He, K., Girshick, R., et al.: Faster R-CNN: towards real-time object detection with region proposal networks. IEEE Trans. Pattern Anal. Mach. Intell. **39**(6), 1137–1149 (2017)
6. Fanger, P.O.: Thermal comfort analysis and applications in environment engineering **2**, 225–240 (1970)
7. Mousavi, S., Rismanchi, B., Brey, S., et al.: Thermal and energy performance evaluation of a full-scale test cabin equipped with PCM embedded radiant chilled ceiling. Build. Environ. **12**(226), 1–13 (2023)

FYPnet: A Road Scene Perception Algorithm for On-Board Chip Deployable Detection and Segmentation Optimization Fusion

Yunpeng Shi, Yongwei Li, Wenqiang Li, and Xin Zhou

Abstract This work proposes an optimized fusion of object detection and image segmentation network for on-board chip deployable road scene perception. Firstly, the adaptive kernel convolution (AKConv) and C2F module are introduced into end-to-end network model to achieve lightweight and efficient feature extraction. Meanwhile, a novel lightweight multi-scale segmentation head module (RPFseg) based on the fusion of YOLO and PSPNet is proposed to enhance the model so as to reduce the size and parameters of the network and improve the computation speed. With the fusion of the improved modules, an optimized network structure (named FYPnet) is designed to focus on discriminative features and improve the quality of feature fusion. Experimental results show that the proposed network owns excellent performances in terms of mean intersection over union (mIoU) and frames per second (FPS).

Keywords Object detection · Image segmentation · Road scene · End-to-end model · On-board chip

Y. Shi · Y. Li · W. Li
Chongqing Changan Automobile Co., Ltd, Chongqing, China
e-mail: shiyp@changan.com.cn

Y. Li
e-mail: liyw3@changan.com.cn

W. Li
e-mail: liwj5@changan.com.cn

X. Zhou (✉)
China Automotive Engineering Research Institute Co., Ltd., Chongqing, China
e-mail: zhouxin@caeri.com.cn

J. S. Park et al. (eds.), *Advanced Multimedia and Ubiquitous Engineering*, Lecture Notes in Electrical Engineering 1475, https://doi.org/10.1007/978-981-95-1565-3_19

1 Introduction

The integration of object detection and image segmentation technologies plays a pivotal role. Object detection facilitates the identification and tracking of diverse elements on the road, such as vehicles, pedestrians, and traffic signs [1]. This capability contributes to a comprehensive understanding of the surrounding environment, enhancing the overall efficacy of road scene perception for intelligent vehicles [2]. Furthermore, image segmentation technology allows the transportation system to partition the road scene into different regions, further enhancing its understanding and responsiveness to complex environments. The combination of the two technologies is widely used intelligent vehicles to achieve safe and efficient autonomous driving.

Recent years, substantial advancements have been seen in the field of object detection technology for road scenes [3]. R-CNN utilizes selective search to generate candidate regions, followed by feature extraction using convolutional neural networks (CNN), and ultimately employs support vector machines (SVM) for target classification. This approach has become emblematic of two-stage detection algorithms. In contrast, YOLO, representing the one-stage detection algorithms, adopts a grid-based strategy, simultaneously performing object detection and classification in each grid of the image. This enables efficient real-time detection by dividing the image into grids and processing them concurrently. Particularly noteworthy is the remarkable performance of YOLO in pedestrian detection tasks, providing robust support for real-time scenarios [4]. Meanwhile, the image segmentation technology in road scenes has also been widely studied [5]. Through advanced methods like deep convolutional neural networks, researchers have successfully segmented road scenes into fine-grained semantic regions, including roads, pedestrian areas, and lane markings [6].

To further improve the performance of the network and achieve on-board chip deployment for object detection and image segmentation functionalities in road scenes, this paper proposes an end-to-end on-board chip deployable road scene perception network. A lightweight and efficient feature extraction module for single-stage object detection is analyzed and adopted. Derived from the structure of pyramid pooling, a novel lightweight head module (named RPFseg) is designed for multi-scale segmentation. With combining the improved modules, an optimized network, named FYPnet, is proposed. Finally, the network is trained and tested on the Cityscapes dataset to verify the performances of mIoU and FPS. Moreover, the proposed network is also deployed on on-board chip to show the application value.

2 Design of the Feature Extraction Module

To achieve the goal of high precision road scene perception, the YOLOv5 backbone is employed and optimized in this work to achieve feature extraction. Since feature extraction plays a crucial role in computer vision tasks and directly influence the understanding and representation ability of input data, CSPDarknet53 is employed as the backbone network in this work. In this network, the feature maps are divided into two parts [7], where the first part directly connects to the next stage and the other one undergoes a series of convolutional operations before connecting. To enhance the capability of the model to comprehend and represent input data effectively, an improved network is designed for backbone section to balance the information flow. Briefly, the frame of the improved network is shown in Fig. 1.

In Fig. 1, P1 ~ *P*5 represents the feature maps with strides of 2 ~ 32 in downstream processing, respectively. These feature maps are utilized for handling detection targets of various scales. Compared with original YOLOv5 backbone, it can be seen in Fig. 1 that the AKConv (Adaptive kernel convolution) and C2F module are introduced. The structure of AKConv is illustrated in Fig. 2, it can be found that AKConv primarily offers a flexible convolution mechanism, allowing the convolution kernel to have an arbitrary number of parameters and sampling shapes. This convolution breaks the limitations of traditional convolution, which is confined to fixed local windows and sampling shapes, enabling convolution operations to more precisely adapt to different datasets and target positions. Meanwhile, the design of flexible convolution kernels allows them to have any number of parameters, facilitating adjustments in size and shape according to specific requirements. On this basis, the accuracy of feature extraction can be improved through adaptive sampling position adjustments. Additionally, AKConv supports linearly increasing or decreasing the number of convolution parameters, aiding in optimizing performance in hardware environments, which is suitable for application with lightweight models. Therefore, AKConv is employed in this network to replace the traditional convolution mechanism. Additionally, the C2F module, which includes the CSP bottleneck structure, improves the extraction of feature information while reducing parameters and computational workload. Therefore, the chunk function is employed to split the output into two parts so as to optimize the distributed computation performance.

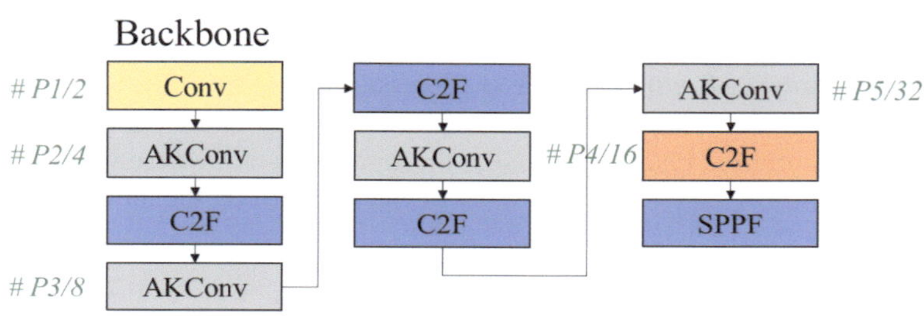

Fig. 1 Frame of the improved network for backbone section

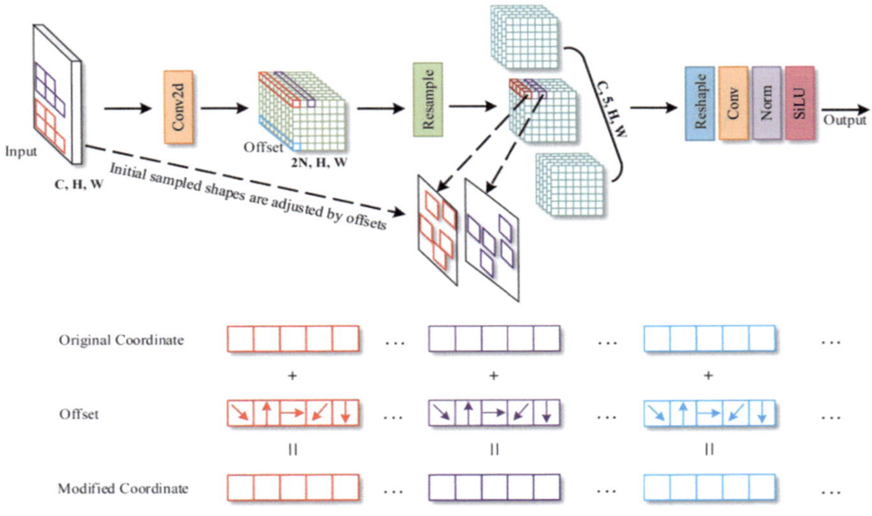

Fig. 2 The structure of AKConv

3 Design of Image Segmentation Head Module

With using the feature extraction module, the corresponding features can be extracted. Then, the head module is designed to achieve image segmentation. In this work, a novel lightweight head module (named RPFseg) is proposed for multi-scale segmentation. The fundamental concept of this module revolves around leveraging the pyramid pooling module to capture contextual information across various scales, thereby augmenting the comprehension of image semantics and improving segmentation accuracy. The frame of the RPFseg module is shown in Fig. 3. Firstly, a concatenation is performed in the initial input stage, merging a low-resolution feature map containing abundant semantic information with a high-resolution feature map that holds less semantic information, thereby creating the input feature map. Subsequently, the feature map undergoes multi-scale information extraction through the RFB2 module, aiding the network in capturing both fine-grained details and global information, thereby enriching the feature representation. Accordingly, the pyramid pooling module concatenates features from each scale of the input feature map, creating a vector fused with features at different scales. Next, the FFM module, incorporating a channel attention mechanism, is applied to adaptively adjust the importance of different channels. Finally, these modules are combined into a unit to emphasize discriminative features, consequently enhancing the quality of feature fusion.

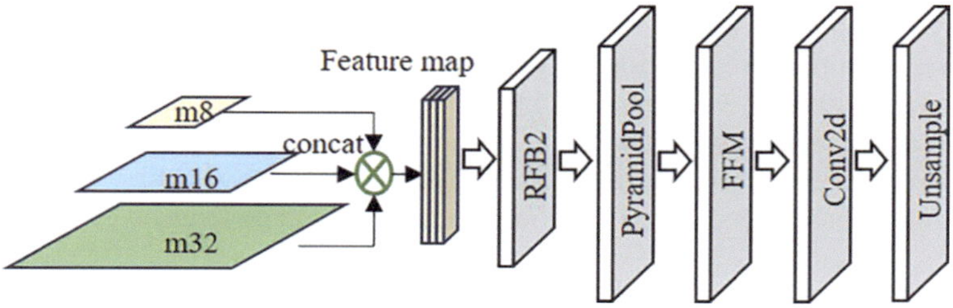

Fig. 3 The network frame of the RPFseg module

4 Structure of the FYPnet Network

By comminating the optimized backbone section and the proposed RPFseg module, the FYPnet (Fusion of YOLO and PSPNet) network is emerged and the structure of this network is illustrated in Fig. 4.

In the optimized network, there are three main parts: backbone, neck and head. In the Backbone section, the AKConv module replaces the traditional convolution mechanism, and the C3 module is substituted with the C2F module to achieve lightweight and efficient feature extraction. The multi-level feature maps generated in the backbone section undergo operations such as upsampling and fusion in the PAFPN structure, resulting in the final feature map used for both object detection and image segmentation. In the head section, the optimized RPFseg image segmentation module integrates the RFB2, Pyramid-Pooling, FFM modules to achieve multi-scene perception. Specifically, this module is designed to improve the performance of the

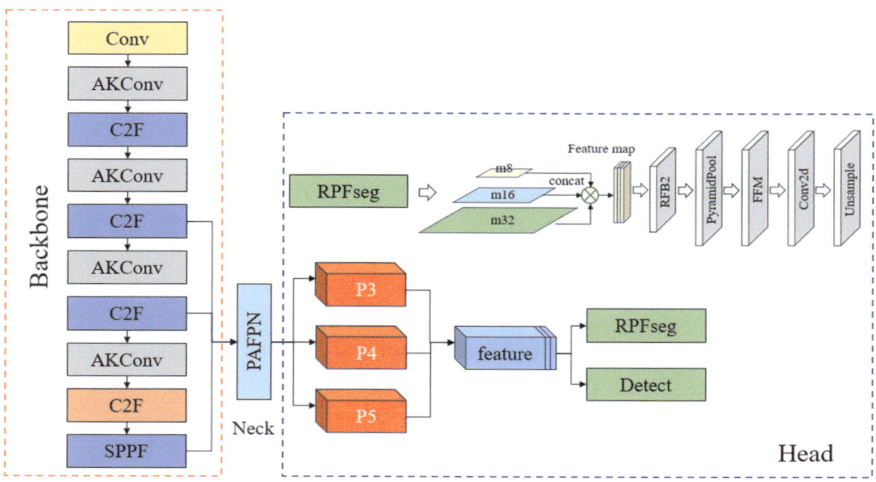

Fig. 4 Frame of FYPnet network

network in capturing both fine-grained details and global information, and finally enriching feature representation. In this work, the FYPnet network is designed to focus on discriminative features and improve the quality of feature fusion. Furthermore, the designed network can achieve training and prediction simultaneously for both object detection and image segmentation. With using the proposed network, less parameters are needed and the computation cost can be reduced.

5 Performance Test and Deployment Test

5.1 Performance Test

To verify the performance of the proposed network model, the Cityscapes dataset [8] is employed for both training and testing in terms of mean intersection over union (mIoU) and frames per second (FPS) indexes. Meanwhile, the other methods, for instance, BiSeNetV2-L, BiSeNetV2-B, DF1-Seg, TinyHMseg, etc., are used to make comparisons. Accordingly, the test results are shown in Fig. 5. It is obvious in Fig. 5 that the proposed network can handle both detection and segmentation tasks, simultaneously. Meanwhile, the proposed FYPnet network obtains the FPS performance index of 82 and the mIoU is 74.4%. Compared to other methods, it can be found that the proposed network has better FPS than those of BiSeNetV2-L, BiSeNetV2-B, DABNet, ICNet, TinyHMseg, FasterSeg and DFANetA networks. Furthermore, the mIoU performance of the FYPnet network only slightly less than those of BiSeNetV2-L and BiSeNetV2-B network. Performance test show that the proposed FYPnet network owns better comprehensive performance than other methods in mIoU and FPS, revealing the effectiveness of the improvement.

5.2 On-Board Chip Deployment Test

To further verify the ability of the proposed FYPnet network for on-board chip deployment, the real running tests are conducted. Dring the tests, the proposed network is deployed and run on automotive-grade chips by converting it to an ONNX model and subsequently transforming it into a BPU model. The specific conversion process is given in Fig. 6. Test results shows that the network can successfully run in the on-board chip platform with a running time of 224 ms, revealing the real-time application value of the proposed network for intelligent vehicles.

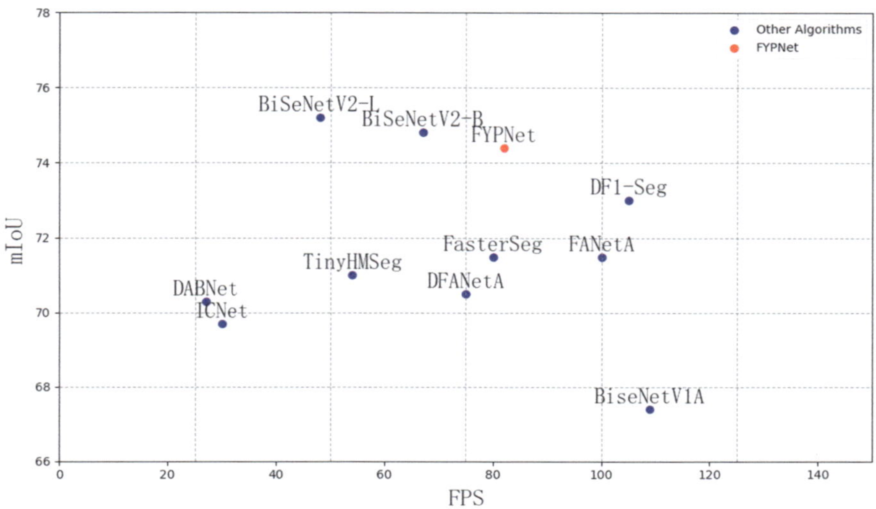

Fig. 5 mIoU and FPS test results comparison

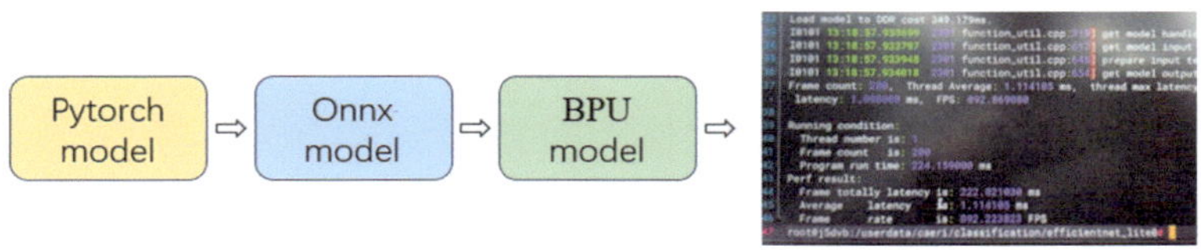

Fig. 6 Model conversion process and deployment

6 Conclusion

In this work, an optimized fusion of object detection and image segmentation network is proposed for road scene perception. AKConv and C2F module are employed in this network model to achieve lightweight and efficient feature extraction. A novel lightweight multi-scale RPFseg module is designed to improve the computation speed. Experimental tests demonstrate the performance of the fusion FYPnet network in mIoU and FPS. The extension of the proposed method can be multi-modules fusion.

Acknowledgements This work was supported by Science and Technology Innovation Key R&D Program of Chongqing (CSTB2023NSCQ-LZX0172)

References

1. Yuan, S., Zhang, Q., Zhu, L., et al.: Multi-level object detection by multi-sensor perception of traffic scenes. Neurocomputing **514**(12), 486–499 (2022)
2. Zhu, H., Yuen, K.V., Mihaylova, L., et al.: Overview of environment perception for intelligent vehicles. IEEE Trans. Intell. Transp. Syst. **18**(10), 2584–2601 (2017)
3. Zhao, Z.Q., Zheng, P., Xu, S.T., et al.: Object detection with deep learning: a review. IEEE Trans. Neural Netw. Learn. Syst. **30**(11), 3212–3232 (2018)
4. Zhang, Q., Wang, Y., Song, L., et al.: Using an improved YOLOv5s network for the automatic detection of silicon on wheat straw epidermis of micrographs. J. Field Robot. **40**(1), 130–143 (2023)
5. Awate, S.P., Tasdizen, T., Whitaker, R.T.: A development of navigation system with image segmentation in mosaicplasty of the knee. In: IEEE International Conference on Fuzzy Systems (2012)
6. He, K., Zhang, X., Ren, S., et al.: Spatial pyramid pooling in deep convolutional networks for visual recognition. IEEE Trans. Pattern Anal. Mach. Intell. **37**(9), 1904–1916 (2015)
7. Zhang, Z., Lu, X., Cao, G., et al.: ViT-YOLO: transformer-based YOLO for object detection. In: International Conference on Computer Vision (2021)
8. Pan, H., Hong, Y., Sun, W., et al.: Deep dual-resolution networks for real-time and accurate semantic segmentation of traffic scenes. IEEE Trans. Intell. Transp. Syst. **24**(3), 3448–3460 (2023)

A Study of Absolute Planar Magnetic-Field-Shared Angular Displacement Sensors

Liu Yongshang, Yang Yaquan, Wang Yang, Tang Qingyun, Luo Jiufei, and Tang Qifu

Abstract An absolute planar magnetic-field-shared angular displacement sensor is proposed in response to the increasing demand for miniaturized sensors with a large ratio of inner and outer diameters in the industrial field. The sensor utilizes PCB (Printed Circuit Board) technology to design the stator part as a structure in which two sets of induction coils share the toroidal magnetic field generated by the inner and outer ring excitation coils, and the rotor part as a planar magnetic-field-shared structure. This paper describes the structure and working principle of the sensor, and simulation experiments are carried out to verify its feasibility. The experimental results show that the error of the sensor is − 46 ~ + 87 "within the pair of poles the error is − 127 ~ + 412" in the range of 0°–360°.

Keywords Absolute angular displacement · Inductive sensors · Double-layer-copper foil array

1 Introduction

Due to the development of manufacturing industry gradually moving towards automation and intelligence, the application scope of closed-loop control system in industrial production is expanding. As a key part of the closed-loop control system, the displacement sensor has more stringent requirements in practical applications. Displacement sensors should not only be able to work in the oil, dust and other working environments, but also to meet the needs of anti-interference ability, reliability and so on [1, 2]. Inductive sensors are suitable for harsh operating environments and have relatively low processing and installation requirements [3].

L. Yongshang · Y. Yaquan · W. Yang · T. Qingyun · T. Qifu (✉)
Chongqing University of Technology, Chongqing, China
e-mail: tqf@cqut.edu.cn

L. Jiufei
Chongqing Univercity of Posts and Telecommunications, Chongqing, China

J. S. Park et al. (eds.), *Advanced Multimedia and Ubiquitous Engineering*, Lecture Notes in Electrical Engineering 1475, https://doi.org/10.1007/978-981-95-1565-3_20

However, the inductive angular displacement sensor is limited by the size of the mechanical structure and the mutual interference between the windings, the current feasible technical conditions to realize the two functions can only be two layers of coaxial sensors coaxial installation. Such two-layer coaxial sensors have large errors in their angular displacement measurements due to the differences in the mechanical structure, windings, and hardware circuits of the two sensors. The planar magnetic-field-shared angular displacement sensor proposed in [4] realizes the improvement of the mechanical structure and the simplification of the winding and hardware circuits by combining the outer stator of the two-layer coaxial sensors into a single stator to provide an excitation magnetic field to be shared to the inner two-layer rotor.

In summary, this paper uses the principle of electromagnetic induction and the method of time-varying magnetic field constraints to propose a planar magnetic-field-shared absolute angular displacement sensor. The sensor not only makes the magnetic flux change into sinusoidal rule of change in two groups of induction coils, but also shares the same excitation magnetic field generated by the same excitation coil, which is suitable for the angular displacement measurement under the working environment where the difference between the inner and outer diameters is small.

2 The Structure and Working Principle of the Absolute Planar Magnetic-Field-Shared Angular Displacement Sensor

2.1 *The Structure of the Absolute Planar Magnetic-Field-Shared Angular Displacement Sensor*

Based on the principle of magnetic field generation and the properties of the electromagnetic inter, through the design of the shape of the excitation coil to constrain the distribution area of the magnetic field, the reasonable use of the shape of the induction coil, so that the magnetic flux change in the magnetic field induction coil and the displacement of the copper foil close to the sinusoidal relationship. In order to obtain the absolute position, the structure of a planar magnetic-field-shared absolute angular displacement sensor is designed, as shown in Fig. 1.

2.2 *The Working Principle of the Absolute Planar Magnetic-Field-Shared Angular Displacement Sensor*

Absolute planar magnetic-field-shared angular displacement sensor adopts "precision machine + precision machine" combination measurement [5] to realize the absolute displacement measurement. The angular displacement of the sensor is calculated

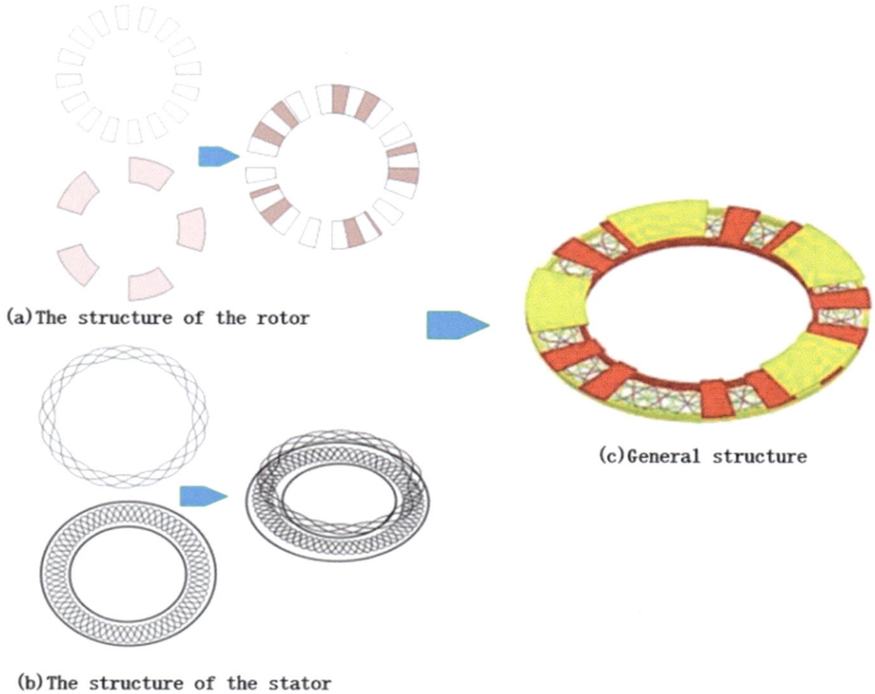

Fig. 1 The structure of the absolute planar magnetic-field-shared angular displacement sensor

by the inverse tangent of the upper and lower sensing signals generated in the above sensor structure. Assuming that one set of output sensing signals of the sensor is $e_{\sin}$ and $e_{\cos}$, the difference of angular displacement $\Delta\theta$ can be derived from Eq. (1).

$$\begin{cases} \Delta\theta = \arctan\left(\frac{e_{\sin}}{e_{\cos}}\right) & (e_{\sin} > 0, e_{\cos} > 0) \\ \Delta\theta = \arctan\left(\frac{e_{\sin}}{e_{\cos}}\right) + \pi & (e_{\sin} > 0, e_{\cos} < 0) \\ \Delta\theta = \arctan\left(\frac{e_{\sin}}{e_{\cos}}\right) + 2\pi & (e_{\sin}\langle 0, e_{\cos}\rangle 0) \\ \Delta\theta = \arctan\left(\frac{e_{\sin}}{e_{\cos}}\right) + \pi & (e_{\sin} < 0, e_{\cos} < 0) \end{cases} \tag{1}$$

Assuming that the number of poles of the two sets of induction coils are N_U and N_L ($N_L < N_U$) and are of mutual mass, the angular displacements induced by the two sets of induction coils when the rotor of the sensor is rotating are $\Delta\theta_U$ and $\Delta\theta_L$, which can be derived from Eqs. (2) and (3)

$$\Delta\theta_U = \Delta\theta\frac{2\pi}{N_U} \tag{2}$$

Table 1 Correspondence between ordinal numbers I and I′

I	I′
0	0
1	5
2	10
3	15
4	4
5	9
6	14
7	3
8	8
9	13
10	2
11	7
12	12
13	1
14	6
15	11

$$\Delta\theta_{\mathrm{L}} = \Delta\theta\frac{2\pi}{\mathrm{N_L}} \tag{3}$$

When the rotor rotates in the full circumference range, $\Delta\theta_{\mathrm{U}}$ repeats $\mathrm{N_U}$ times and $\Delta\theta_{\mathrm{L}}$ repeats $\mathrm{N_L}$ times, which can be seen from Eq. (4).

$$\begin{cases} I = \mathrm{N_U}\mathrm{N_L}(\Delta\theta_{\mathrm{L}} - \Delta\theta_{\mathrm{U}}) & (\Delta\theta_{\mathrm{L}} \geq \Delta\theta_{\mathrm{U}}) \\ I = \mathrm{N_U}\mathrm{N_L}\left(\Delta\theta_{\mathrm{L}} - \Delta\theta_{\mathrm{U}} + \frac{2\pi}{\Delta\theta_{\mathrm{L}}}\right) & (\Delta\theta_{\mathrm{L}} \leq \Delta\theta_{\mathrm{U}}) \end{cases} \tag{4}$$

The angular displacement sensing diagonal position θ usually consists of two parts, an integer part and a fractional part. In this case, $\Delta\theta_{\mathrm{U}}$ is the fractional part and the ordinate I' is the integer part. However, the actual ordinal number I' does not indicate when the absolute angular position θ changes from 0 to 2π from small to large, and needs to be combined with a logarithmic internal equivalent $2\pi/\mathrm{N_U}$. The relationship between the angular position θ and the actual ordinal number [6–8] can be derived from Eq. (5) and the relationship [9, 10] between I and I' is shown in Table 1.

$$\theta = \frac{2\pi I'}{\mathrm{N_U}} + \Delta\theta_{\mathrm{U}} \tag{5}$$

3 Model Simulation and Analysis

The simulation of the absolute planar magnetic-field-shared angular displacement sensor adopts the number of cycles of the induction coil as 5 and 16 respectively, and the simulation time of the sensor model is 500 ns, the double-layer copper foil rotor of the sensor rotates 72 degrees, and the simulation step is set to 1.8 degrees, and the simulation is carried out by taking 40 rotor positions, and the results of the simulation are shown in Figs. 2 and 3, in which the standing wave induction signal of the 5-pole induction coil sets are shown in Fig. 2 and the standing wave induction signal of the 16-pole induction coil sets are shown in Fig. 3.

The simulation results show that there is a specific relationship between each rotational position of the sensor rotor and the induced electromotive force of the induction coil. Functionally, the signals from the 5-pole induction coil set determine the approximate relative position of the sensor between the stator and rotor, while the signals from the 16-pole induction coil set ensure the accuracy of the sensor.

The sensor angular displacement signal can be solved by performing an inverse tangent operation on the two sets of signals. In addition, 16-pole induction coil set generates a stronger signal than the 5-pole induction coil set because the 16-pole induction coil set is located closer to the position of the rotor.

And the calculated curves for 5-pole and 16-pole induction coil set are placed in the same graph as shown in Fig. 4. In one simulation cycle, the electric potential of

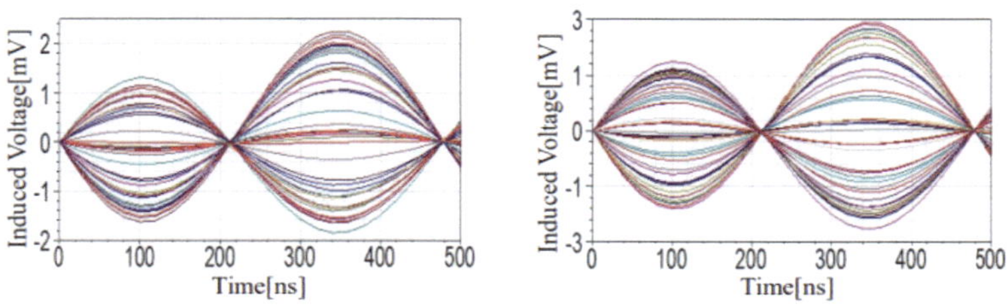

(a) Sine standing wave (b) Cosine standing wave

Fig. 2 Simulation of the 5-pole induction coil (*each curve represents a rotor position*)

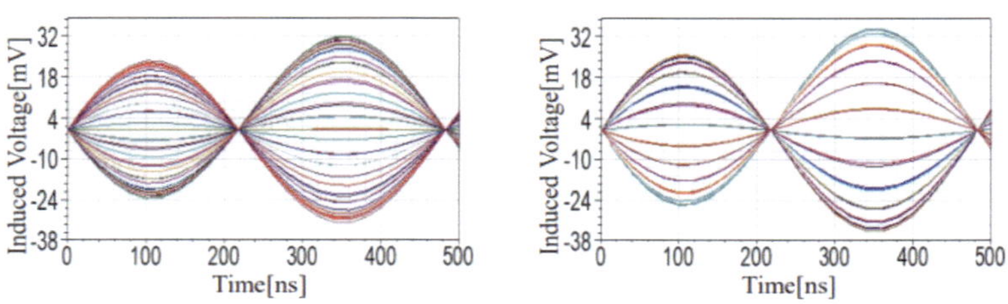

(a) Sine standing wave (b) Cosine standing wave

Fig. 3 Simulation of the 16-cycle induction coil (*each curve represents a rotor position*)

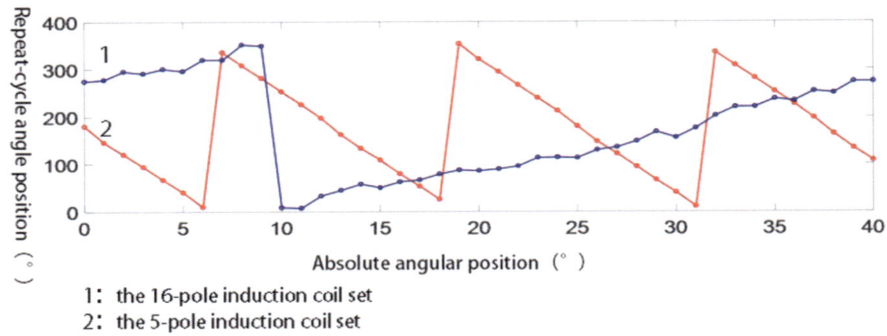

Fig. 4 The absolute position of the rotor of the angular displacement sensor (*5&16 poles*)

the 5-pole induction coil set changes by one cycle, while the 16-pole induction coil set changes by 3.2 cycles. From this, the relationship between the positions of the 5-pole and 16-pole induction coil set can be determined, thus realizing the absolute measurement, which is in line with the theoretical derivation in the previous section.

4 Conclusion

In this paper, an absolute planar magnetic-field-shared angular displacement sensor is proposed, using PCB (Printed Circuit Board) technology to design the stator part as a structure with two sets of induction coils sharing the toroidal magnetic field generated by the inner and outer ring excitation coils, and the rotor part as a structure with a double-layer copper foil array. The sensor can be miniaturized and is suitable for applications with a large ratio of inner and outer diameters. However, there is a limitation in the structure. The accuracy of the sensor is higher only if the distance between the stator and rotor is within a certain range. The operating principle and structural feasibility of the sensor are verified by model simulation.

References

1. Gao, F., Guo, W., Song, Q.: Current development of heavy-duty manufacturing equipments. J. Mech. Eng. **19**, 92–107 (2010)
2. Lin, C.: Situation analysis and prospect of grating encoders. J. World Manuf. Eng. Market **01**, 101–106 (2014)
3. Dinulovic, D., Hermann, D., Fluegge, J.: Development of a linear micro-inductosyn sensor. J. Magnetics **10**, 2830–2832 (2006)
4. Luo, P.: Design and Experimental Study of Angular Displacement Sensor with Shared Magnetic Field. D. Chongqing University of Technology (2020)
5. Tang, Q.: Study on Time Grating Displacement Sensor Based on Precise Confine Method of Time-varying Magnetic Field. D. Chongqing University (2015)

6. Gu, X.: Study on the Measurement Method and Sensor of Absolute Linear Displacement by Bilateral Sensing Units. D. Chongqing University of Technology (2020)
7. Wu, L.: Research on the Double Track Absolute Encoder and Experiment Device. D. Beijing Jiaotong University (2015)
8. Zhou, W.: Time-grating Angular Displacement Sensor Self-calibration. D. Chongqing University of Technology (2012)
9. Wang, Y., Tang, Q., Chen, X.: Absolute time-grating displacement sensor realization method based on couple of multi-pole combination. J. Instrum. Tech. Sensor **04**, 20–22 (2017)
10. Gao, W., Shi, H., Zhou, H.: Research on a Novel Signal Processing Technique for Planar Inductive Angle Sensors. J. Chongqing Univ. Technol. **11**, 177–183 (2020)

Integrated Automatic Parking Path Planning and Trajectory Tracking Optimization Method

Changhao Piao, Yongkang Su, and Junren Shi

Abstract Path planning and trajectory tracking control in current automatic parking systems are usually independent of each other, resulting in more complex computations. Therefore, this paper proposes an integrated automatic parking path planning and trajectory tracking optimization method based on Multi-stage Nonlinear Model Predictive Control (MSNMPC). Each phase defined by the MSNMPC controller has specific cost and constraint functions that depend only on the vehicle state and control inputs of that phase to satisfy the constraints of all phases of the parking process. In addition, the method integrates path planning and tracking control into a single optimization problem, which is solved online to achieve integrated parking control. By comparing with the hierarchical control architecture based on MPC and RRT*, the proposed integrated parking scheme has good flexibility and tracking performance, and the parking elapsed time is reduced by 10.47%.

Keywords Automobile engineering · Automatic parking · MSNMPC · Path planning · Trajectory tracking

1 Introduction

Currently, automated parking control typically uses a hierarchical control method that separates the path planning and trajectory monitoring parts of the automatic parking issue into two levels. Parking path planning in the higher layer is frequently done using techniques like search-based methods [1], polynomial approaches [2], and B-spline curves [3]. Algorithms like fuzzy control [4], neural networks [5], and Model Predictive Control (MPC). Qin [6] proposed a path planning method for automated valet parking (AVP) systems based on directional graph search and

C. Piao · Y. Su · J. Shi (✉)
School of Automation, Chongqing University of Posts and Telecommunications, Chongqing, China
e-mail: shijr@cqupt.edu.cn

C. Piao
e-mail: piaoch@cqupt.edu.cn

J. S. Park et al. (eds.), *Advanced Multimedia and Ubiquitous Engineering*, Lecture Notes in Electrical Engineering 1475, https://doi.org/10.1007/978-981-95-1565-3_21

geometric curves, using an improved C-type vertical parking path planning algorithm to generate parking paths, which provides flexibility in the selection of parking start nodes. Wang [7] proposed a path tracking method based on model predictive control and variable prediction horizon, analyzed the response of the path tracking control system under different prediction horizon, and designed a particle swarm optimization algorithm to achieve the adaptive optimization of prediction horizon. An enhanced RRT * algorithm with Reeds–Shepp curves and nonlinear optimization were combined to create a hierarchical planning algorithm by Wang [8]. This algorithm can plan an efficient parking trajectory with continuous curvature under various starting positions and parking scenarios. Li [9] suggested combining tracking control and path quadratic planning into a single MPC optimization problem, which would then be solved by increasing the MPC cost function by the collision risk cost. This approach doesn't solve the parking problem in all scenarios because it still depends on the assumption that an optimal path can be found beforehand. Nevertheless, the same issue is present in the previously mentioned studies, which give the vehicle dynamics, kinematics, and constraints insufficient consideration [10, 11]. Furthermore, because of the interaction between the upper and lower control layers, the hierarchical control scheme used in these studies may produce planned paths that are not sufficiently smooth or reasonable, making it impossible to successfully complete the parking task in some complex parking environments.

The parking control scheme proposed in this paper represents the evolution of uncertainty parameters through a scenario tree [12], and the MSNMPC controller defines a specific cost and constraint function for each stage, which relies only on the vehicle state and control inputs at that stage. Meanwhile, the path planning and tracking control are integrated into one optimization problem, and the computational efficiency is improved by online solving to ensure the optimality of the parking path and trajectory tracking.

2 Control System Design

2.1 Multi-stage NMPC

Using the MSNMPC-based control scheme to design the controller, the uncertainty that exists during the parking process can be modeled with a scenario tree, as shown in Fig. 1. In addition, MSNMPC is the best solution for the robust NMPC problem [13], because it computes the optimal closed-loop feedback policy within a limited prediction range.

Due to the discrete setting of the scenario tree, the discrete time formulation of an uncertain nonlinear system with uncertain parameters in the scenario tree setting is expressed as follows:

$$\boldsymbol{x}_{k+1}^{j} = f\left(\boldsymbol{x}_{k}^{p(j)}, \boldsymbol{u}_{k}^{j}, \boldsymbol{d}_{k}^{r(j)}\right) \tag{1}$$

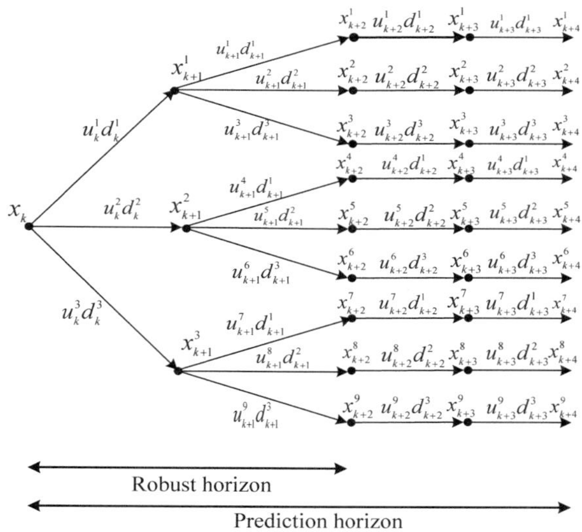

Fig. 1 Scenario tree schematic for multi-stage NMPC uncertainties

where x^j_{k+1} denotes the state variable of the jth node on the scene tree at the $k + 1$ sampling moment, $\boldsymbol{x}^{p(j)}_k$ denotes the amount of the vehicle state of the $p(j)$ parent node at the k sampling moment, $\boldsymbol{u}^j_k$ denotes the control input of the jth node at the k sampling moment, $\boldsymbol{d}^{r(j)}_k$ denotes the uncertainty parameter at the k sampling moment.

The built scenario tree model indicates that when the prediction horizon is expanded, the optimization problem's size tends to increase exponentially [14]. We can presume that following a particular robust horizon, the uncertainty will stay constant. The optimal control problem for the scenario-based multistage formulation at each time step k can be expressed as follows:

$$\min_{\boldsymbol{x}^j_k, \boldsymbol{u}^j_k} \sum_{i=1}^{S} \omega_i J_i\left(\boldsymbol{x}^j_{k+1}, \boldsymbol{u}^j_k\right), \forall (j, k) \in \boldsymbol{I} \tag{2}$$

$$\begin{aligned} s.t.\, \boldsymbol{x}^j_{k+1} &= f\left(\boldsymbol{x}^{p(j)}_k, \boldsymbol{u}^j_k, \boldsymbol{d}^{r(j)}_k\right), \forall (j, k) \in \boldsymbol{I} \\ \boldsymbol{x}^j_k &\in \boldsymbol{X}, \forall (j, k) \in \boldsymbol{I} \\ \boldsymbol{u}^j_k &\in \boldsymbol{U}, \forall (j, k) \in \boldsymbol{I} \end{aligned} \tag{3}$$

$$\text{If } \boldsymbol{x}^{p(j)}_k = \boldsymbol{x}^{p(l)}_k, \text{ then } \boldsymbol{u}^j_k = \boldsymbol{u}^l_k, \forall (j, k), (l, k) \in \boldsymbol{I} \tag{4}$$

where $\boldsymbol{I}$ denotes all index combinations (j, k) corresponding to the nodes covered by the scene tree, S denotes the number of scenes, ω_i denotes the weight corresponding to each scene S_i, $\boldsymbol{x}^{p(j)}_k$ and $\boldsymbol{x}^{p(l)}_k$ denote the states with the same parent node at the moment k, $P(j)$ and $P(l)$ denote the two nodes with the same parent node, j and

l respectively, $\boldsymbol{u}_k^j$ and $\boldsymbol{u}_k^l$ denote the control inputs with the same parent node, and $J_i\left(\boldsymbol{x}_{k+1}^j, \boldsymbol{u}_k^j\right)$ denotes the cost function of each scene S_i. The definitions are as follows:

$$J_i\left(\boldsymbol{x}_{k+1}^j, \boldsymbol{u}_k^j\right) = \sum_{k=0}^{N_p-1} L\left(\boldsymbol{x}_{k+1}^j, \boldsymbol{u}_k^j\right), \forall\left(\boldsymbol{x}_{k+1}^j, \boldsymbol{u}_k^j\right) \in S_i \tag{5}$$

where v_{ref} is the desired vehicle longitudinal velocity and $\Delta \boldsymbol{u}$ is the optimal open-loop control sequence.

2.2 System Constraints and Optimization Problem Solving

The safety constraints during parking are designed to constrain the lateral position and longitudinal position of the vehicle during parking as follows:

$$\begin{aligned} &Y_{down} + b \leq \boldsymbol{x}(k_s + 1) \leq Y_{up} - a \\ &\boldsymbol{x}(k_s + 1) \geq X_{bound} + w/2 \\ &i = 1, 2, 3, \ldots, N_p \\ &dist_{\min} \geq dist_{safe} \end{aligned} \tag{6}$$

where Y_{up} and Y_{down} are the upper and lower boundary positions of the parking space respectively, a is the sum of the front overhang length and wheelbase as well as the safety margin of the front end, b is the sum of the rear overhang length and the safety margin of the rear end, X_{bound} is the side boundary of the garage, w is the width of the vehicle, $dist_{min}$ denotes the minimum distance from the vehicle to the obstacle, and $dist_{safe}$ denotes the parking safety distance.

The integrated automatic parking path planning and trajectory tracking optimization method proposed in this paper integrates vehicle speed planning, path planning and tracking control into a single optimization problem to obtain a description of the control problem that obeys the described prediction model and the described constraints as follows:

$$\min_{\Delta \boldsymbol{u}(k)} J(\boldsymbol{x}(k_s + i), \Delta \boldsymbol{u}(k_s + i - 1)) \tag{7}$$

Further, the interior point method is used to solve the said control problem to obtain the optimal open-loop control sequence $\Delta \boldsymbol{u}$, and the first column of elements in the optimal open-loop control sequence is used to control the vehicle to realize automatic parking, and the optimal open-loop control sequence $\Delta \boldsymbol{u}$ is denoted as:

$$\Delta \boldsymbol{u} = \arg\min_{\Delta \boldsymbol{u}} \left\{ \sum_{i=1}^{N_P} \left[\left(v(k_s+i) - v_{ref}(k_s) \right)^2 \cdot \boldsymbol{Q} \right] + \sum_{i=1}^{N_P} \left[\Delta \boldsymbol{u}(k_s+i-1)^2 \cdot \boldsymbol{R} \right] \right\} \tag{8}$$

where v_{ref} is the desired vehicle longitudinal velocity and $\Delta \boldsymbol{u}$ is the optimal open-loop control sequence.

3 Simulation and Result Analysis.

In this paper, the proposed MSNMPC integrated control scheme based automatic parking path planning and trajectory tracking optimization method is designed to be validated in an oblique parking scenario based on the MATLAB simulation platform and compared with the hierarchical control scheme based on RRT* and MPC [15].

As shown in Fig. 2, the integrated control scheme exhibits higher flexibility throughout the parking process, with the parking route more closely matching the planned path, while the driving inflection points are also closer to the inflection points of the planned path. The parking time of the integrated control scheme is 39.8 s, which is 10.47% less than the 44.6 s of the comparison experiment, and achieves the requirement of improving parking efficiency. At the same time, the safety distance of the vehicle entry stage is larger compared to the comparison experiment, which improves the safety of the parking process.

Figure 3 illustrates the attitude of the vehicle during the parking process under two different scenarios. Where (a) and (b) represent the real-time position of the vehicle

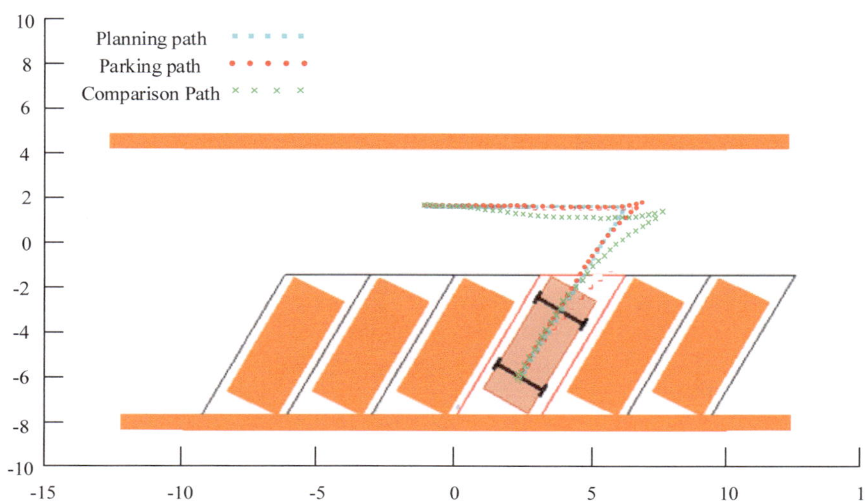

Fig. 2 Simulation of automatic parking process

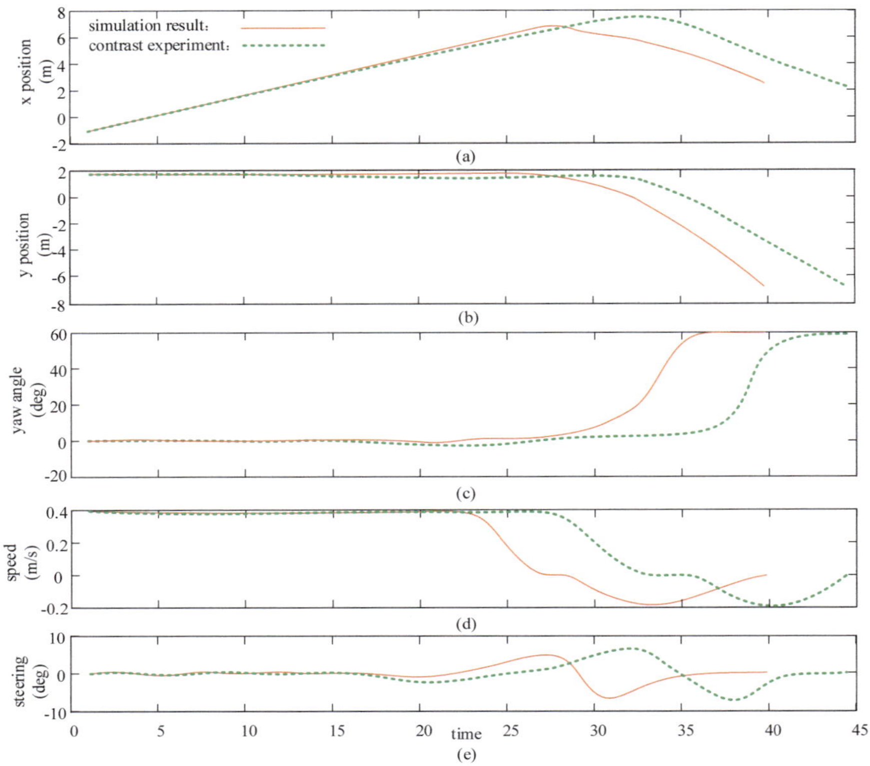

Fig. 3 Real-time vehicle status during simulation

during the automatic parking process and (c) represents the real-time heading angle of the vehicle. The results show that the integrated control scheme enables the vehicle to enter reverse parking earlier and reach the target position earlier. Figures (d) and (e) show that when the vehicle passes through the parking inflection point, the vehicle of the hierarchical control scheme stays longer at the inflection point, which leads to an unsmooth parking process and consequently a longer parking time. Meanwhile, the comparison experimental paths deviate farther from the planning path, i.e., the lateral error is larger.

4 Discussion

The integrated automatic parking path planning and trajectory tracking optimization method based on MSNMPC proposed in this paper solves the problem that the traditional automatic parking system can not complete the task quickly and efficiently in complex environments. The method integrates path planning and trajectory tracking into one optimization problem in the parking process, which obtains a better parking path, improves the parking efficiency and saves time. Through the establishment of scene tree, the uncertainty factor is taken into account in the parking process, which makes the system realize smooth and efficient parking while satisfying the kinematics and system constraints at each stage.

Acknowledgements This work was supported under The National Key Research and Development Program of China (grant number: No.2022YFE0101000)

References

1. Yu, Z., Yuan, J., Li, Y., Yuan, C., Deng, S.: A path planning algorithm for mobile robot based on water flow potential field method and beetle antennae search algorithm. Comput. Electr. Eng. **109**, 108730 (2023)
2. Kou, Y., Ma, C.: Dual-objective intelligent vehicle lane changing trajectory planning based on polynomial optimization. Physica A **617**, 128665 (2023)
3. Li, Z., et al.: Real-time local path planning for intelligent vehicle combining tentacle algorithm and b-spline curve. IFAC-PapersOnLine **54**, 51–58 (2021)
4. Nari, M.I., Mustain, Z., Kautsar, S., Utomo, S.B.: Parallel parking system design with fuzzy logic control. JNTE (2021). https://doi.org/10.25077/jnte.v10n2.904.2021
5. Chai, R., et al.: Design and implementation of deep neural network-based control for automatic parking maneuver process. IEEE Trans. Neural Netw. Learn. Syst. **33**, 1400–1413 (2022)
6. Qin, Z., Chen, X., Hu, M., Chen, L., Fan, J.: A novel path planning methodology for automated valet parking based on directional graph search and geometry curve. Robot. Auton. Syst. **132**, 103606 (2020)
7. Wang, H., Wang, Q., Chen, W., Zhao, L., Tan, D.: Path tracking based on model predictive control with variable predictive horizon. Trans. Inst. Meas. Control. **43**, 2676–2688 (2021)
8. Wang, J., Li, J., Yang, J., Meng, X., Fu, T.: Automatic parking trajectory planning based on random sampling and nonlinear optimization. J. Franklin Inst. **360**, 9579–9601 (2023)
9. Li, S., Li, Z., Yu, Z., Zhang, B., Zhang, N.: Dynamic trajectory planning and tracking for autonomous vehicle with obstacle avoidance based on model predictive control. IEEE Access **7**, 132074–132086 (2019)
10. Lee, S.-H., Chung, C.C.: Autonomous-driving vehicle control with composite velocity profile planning. IEEE Trans. Control Syst. Technol. **29**, 2079–2091 (2021)
11. Li, J., Ran, M., Wang, H., Xie, L.: MPC-based unified trajectory planning and tracking control approach for automated guided vehicles, pp. 374–380 (2019). https://doi.org/10.1109/ICCA.2019.8899955
12. Elorza Casas, C.A., Valipour, M., Ricardez Sandoval, L.A.: Multi-scenario and multi-stage robust NMPC with state estimation application on the Tennessee-Eastman process. Control. Eng. Pract. **139**, 105635 (2023)
13. Lucia, S., Subramanian, S., Limon, D., Engell, S.: Stability properties of multi-stage nonlinear model predictive control. Syst. Control Lett. **143**, 104743 (2020)
14. Holtorf, F., Mitsos, A., Biegler, L.T.: Adaptive Scenario Generation for Multistage NMPC with Shrinking Horizons. IFAC-PapersOnLine **52**, 586–591 (2019)
15. Shi, J., et al.: Planning the trajectory of an autonomous wheel loader and tracking its trajectory via adaptive model predictive control. Robot. Auton. Syst. **131**, 103570 (2020)

A Federated Learning Algorithm Based on Multi-objective Optimization

Daoqu Geng, Shouzheng Wang, and Yihang Zhang

Abstract Federated learning (FL) can achieve secure sharing of data, where all parties participate in model training locally and upload it to the server for aggregation. The data never leaves the parties involved, thus solving the problems of data privacy and data silos. However, FL faces issues such as high communication costs, imbalanced performance distribution among participants, and low privacy protection. To achieve a balance between model accuracy, communication cost, fairness, and privacy, this paper proposes a multi-objective optimization-based federated learning algorithm (M-FedAvg). The multi-objective optimization problem of maximizing the accuracy of the global model, minimizing the communication cost, minimizing the variance of the accuracy and minimizing the privacy budget is solved by NSGA-III. The experimental results show that the algorithm proposed can effectively reduce the communication cost of FL and achieve privacy protection for participants without affecting the accuracy of the global model.

Keywords Federated learning · Multi-objective optimization · NSGA-III · Differential privacy · Communication cost

1 Introduction

With the development of sensor and artificial intelligence, the processing and analysis of data are becoming more and more intelligent. However, the private data generated by the processing can cause huge security risks if uploaded to the cloud. With information security included in key protection [1], researchers often can only analyze data in related fields, leading to the problem of data silos.

FL solves the above problem to some extent [2]. However, Federated Average Algorithm (FedAvg) performs poorly in terms of model accuracy, communication cost, fairness, and privacy. Geyer et al. [3] applied differential privacy (DP) to FL by adding noise to hide whether the participants are involved in training or not. However,

D. Geng (✉) · S. Wang · Y. Zhang
Chongqing University of Posts and Telecommunications, Chongqing, China
e-mail: gengdq@cqupt.edu.cn

J. S. Park et al. (eds.), *Advanced Multimedia and Ubiquitous Engineering*, Lecture Notes in Electrical Engineering 1475, https://doi.org/10.1007/978-981-95-1565-3_22

this brings the problem of balancing model accuracy and privacy [4]. Qolomany et al. [5] optimize the neural network hyperparameters using particle swarm algorithm to achieve single-objective optimization of communication cost. Zhu et al. [6] introduce the Sparse Evolutionary Training (SET) algorithm into the field of FL. Removing the weights with the smallest amount of partial updates at each epoch and optimizing the model hyper-parameters and improving the connectivity parameters of the SET algorithm by the second generation of the Non-dominated Sorting Genetic Algorithm (NSGA-II), thus achieving the minimization of the global model error rate and the communication cost. Zhong et al. [7] used the third generation of the Non-dominated Sorting Genetic Algorithm (NSGA-III), introducing the fairness metrics, and expanding the target space to three dimensions. Meanwhile, Hu et al. [8] take the privacy budget of DP into account and propose to remove weights only at the last epoch, but it will lead to too little removal. Therefore, the proportion of epochs where weights are removed is set to p, for parameter tuning.

This paper takes full advantage of the good performance of the NSGA-III in high-dimensional space, and defines FL as a four-objective optimization problem that maximizes the model accuracy, minimizes the communication cost, minimizes the variance of the accuracy distribution, and minimizes the privacy.

2 Multi-objective Federated Average Algorithm

2.1 Objective Function

The objective functions are: minimizing the global model error rate f_1, minimizing the variance of accuracy f_2, minimizing the communication cost f_3, and minimizing the privacy budget f_4. f_2 and f_4 can be regarded as indicators of fairness and privacy.

$$f_1 = 1 - A = 1 - \sum\nolimits_{k=1}^{K} a_k / K \tag{1}$$

$$f_2 = \sum\nolimits_{k=1}^{K} (a_k - A)^2 / K \tag{2}$$

$$f_3 = \Omega \cdot K \cdot C / K = \Omega \cdot C \tag{3}$$

$$f_4 = \sigma^{-1} \Delta s \sqrt{2CTln(1/\delta)} \tag{4}$$

where A is the global accuracy, and $\{a_1, a_2, \ldots a_K\}$ is the accuracy of the K participants. Ω is the number of participants in FL, and C is the participation ratio. T is the total number of communication rounds in the FL, δ is the relaxation term for DP, σ is the standard deviation of the Gaussian, and Δs is the sensitivity, $\Delta s = \max\left\{\Delta s^{\mathcal{D}_k}\right\} = 2S/m$ for local data of size at least m.

2.2 Decision Variables and Algorithm Flow

The decision variables of the NSGA-III algorithm include the MLP model hyperparameters, the proportion of participation in FL C, the connectivity parameters α and β of the SET algorithm (which are used to control the sparsity of the network), and the standard deviation of the Gaussian distribution σ. The encoding and the parameter meanings of the decision variable in MLP are shown in Fig. 1. The pseudo-code for solving the quad-objective problem is shown in Table 1.

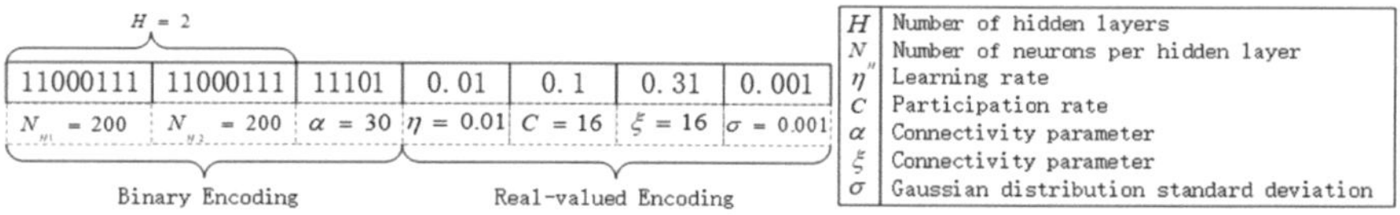

Fig. 1 MLP hybrid encoding

Table 1 The proposed M-FedAvg algorithm

```
Algorithm 1 M-FedAvg Algorithm
Input: Training rounds T, Standard deviation σ, Local training batch size B, Local training
sessions E, Learning rate η, Connectivity parameter α and ξ, subset of K clients S_t
Output: Aggregation model W
Initialize parameter w_0 and primordial population P_n, generate reference points
for population size n ∈ (1,2,...,N) do
    Combine P_n and offspring population Q_n by crossover and mutation
    for individuals i ∈ R_n do
        Initialize a global sparse model w_t^i based on decision variables
        for training rounds t ∈ (1,2,...,T) do
            for client k ∈ S_t parallel do
                for local training sessions e ∈ E do
                    Train the model and remove weights from the posterior p part of B
                    Updated model w_t^k = w_t^k − ∂∇L(w_t^k, b) / max(1, ||g||_2 / S) + N(0, σ_t^2)
                end for
            end for
            Model aggregation w_{t+1}^{avg} = Σ_{k=1}^{K} w_t / K
        end for
        Compute f^i ← (f_1^i, f_2^i, f_3^i, f_4^i) using the global model w_{T+1}^i
    end for
    f = ∪_{i=1}^{2M} f^i, select P_{n+1} by dominated sorting
end for
```

3 Experimental Analysis

3.1 Experimental Setup

Due to limited computational resources, the M-FedAvg is only performed on MLPs with simpler models. The IID distribution and non-IID distribution of the MNIST dataset are used for the dataset. batch size is set to 50, epoch is set to 5, and the parameters of the neural network are shown in Fig. 1. The total number of clients in the FL is $K = 100$, and the proportion of participation is $C = 1$. and the population size and the number of iterations of the NSGA-III are set to 20. The crossover and variance probabilities are set to 0.9 and 0.1. The number of communication rounds required for fitness evaluation in NSGA-III is set to 5 and 10 on the IID and non-IID distributions, respectively. The percentage of deletion p for the improved SET algorithm is set to 10%. The standard deviation of the Gaussian noise is set to 0.001 ~ 0.5 to satisfy the need for DP. The relaxation term is usually smaller than the inverse of the training set size, and thus $\delta = 10^{-4}$ is set.

3.2 Comprehensive Performance Analysis

The standard deviation of the Gaussian distribution is set to 0.05, 0.1, and 0.2, respectively, and the number of communication rounds for FL training is set to 150. It can be seen from Fig. 2 that the accuracy of the global model of the MLP decreases with the increase of the standard deviation of the Gaussian distribution, which verifies that the adjustment can change the degree of privacy preservation, and the addition of a moderate amount of noise does not affect the model's convergence ability.

The number of communication rounds is set to 200 for the experiment of multi-objective optimization of FL. The Pareto optimal solution set is obtained as shown in

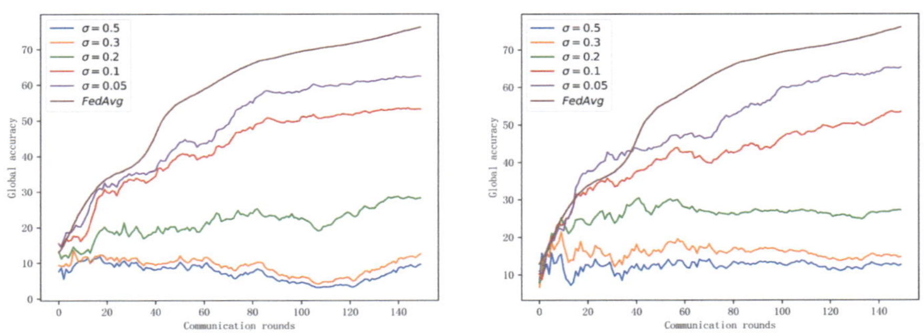

Fig. 2 The relationship between the accuracy of MLP on the IID (left) and non-IID (right) distributions of the MNIST dataset and the standard deviation of the Gaussian distribution

Fig. 3 where the three axes represent the communication cost, accuracy, and variance of accuracy, respectively. The privacy budget is then represented by the color, and the darker the color, the worse the privacy performance. The results show that points with high accuracy tend to have higher communication cost, higher variance of accuracy, or larger privacy budget, indicating that the obtained Pareto solution can effectively achieve the balance of the four objectives.

A solution with high global accuracy is chosen to be noted as Solution 1, and then an inflection point solution is chosen to be noted as Solution 2, and comparison experiments are conducted with the solutions with high accuracy obtained from literature [7] and literature [8] (noted as Solution F1 and F2), as well as the standard solution using full connectivity and no DP (noted as Solution S). The results are shown in Table 2.

Based on the data in the table, the performance of the four-objective optimization method proposed can be discussed in the following four aspects:

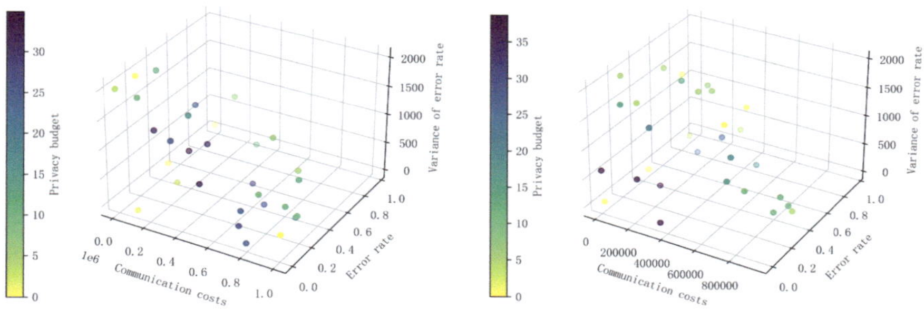

Fig. 3 The Pareto optimal solution set

Table 2 Comparative experiment data of MLP non-IID solutions obtained in MNIST dataset

Parameters	Solution 1	Solution 2	Solution F1	Solution F2	Solution S
Participation rate	0.5896	0.3649	0.6200	0.5866	1
Learning rate	0.1756	0.1668	0.1125	0.1810	0.1
Connectivity Parameters	78	86	95	82	/
Standard deviation	0.2568	0.2568	/	0.3598	/
Hidden layer neurons	0.1509	0.0987	/	0.0603	/
Communication cost	[83]	[78]	[173]	[75]	[200, 200]
Accuracy	36,985	9036	57,553	24,643.06	199,210
Variance of accuracy	97.36	90.33	96.70	92.08	95.69
Privacy budget	3.83	10.69	4.71	/	9.86
Participation rate	2.54	3.68	/	3.46	/

Communication cost: the communication cost of Solution 2 only accounts for about 15.7% of Solution F1, about 36.7% of Solution F2, and about 4.54% of Solution S, which demonstrates that the method proposed can effectively reduce the cost.

Accuracy: the accuracy of Solution 1 is about 0.66% higher than Solution F1, about 5.28% higher than Solution F2, and about 1.67% higher than Solution S, which reflects the improvement of this paper's method in terms of accuracy.

Fairness: Solution F2 and Solution S do not take fairness into account, and the variance of accuracy of Solution 1 is smaller than that of Solution F1, which indicates that the M-FedAvg proposed can improve the fairness of FL training.

Privacy: Solution F1 and Solution S do not consider the problem of participant privacy protection, and the Pareto solution selected can achieve $(3.11, 10^{-4})$-localized DP on average, which is more privacy-protective than the $(3.46, 10^{-4})$-localized DP of solution F2, and it does not seriously affect the quality of the global model.

4 Conclusion

To address the problems of high communication cost, unbalanced performance distribution of participants and poor privacy protection ability of FL, this paper proposes M-FedAvg, which reduces the communication cost by improving the SET algorithm, enhances the privacy protection ability by the DP algorithm, and finally realizes the balance of accuracy, communication cost, fairness, and privacy by adopting the NSGA-III algorithm. Upon experimental results, it is shown that the method proposed can achieve $(3.11, 10^{-4})$-localized DP on the MNIST dataset, and drastically reduces the communication cost, some of the solutions are also improved in accuracy, thus verifying the effectiveness of the method proposed.

However, the method proposed cannot protect against model attacks such as single-point attacks and poisoning attacks. How to improve the security and robustness of the global model is a direction worthy of in-depth research for FL. The decentralized approach through blockchain is a feasible solution. In addition, the computational cost of a multi-objective optimization algorithm is also a direction worth optimizing.

Acknowledgements This work was sponsored in part by the National Key R&D Program of China (No.2022YFE0114300), and in part by the Natural Science Foundation of Chongqing, China (No.cstc2021jcyj-msxmX0330).

References

1. Chen, Z.: Strengthening network security awareness and practicing the requirements of the rule of law: three years of practice and reflection on the cybersecurity law. China Inf. Secur. **06**, 52–55 (2020)
2. McMahan, B., Moore, E., Ramage, D., et al.: Communication-efficient learning of deep networks from decentralized data. In: Artificial Intelligence and Statistics, pp. 1273–1282. PMLR, Amsterdam, Netherlands (2017)
3. Geyer, R.C., Klein, T., Nabi, M.: Differentially private federated learning: a client level perspective. arXiv preprint arXiv:1712.07557 (2017)
4. Tang, L., Chen, Z., et al.: Research progress of privacy issues in federated learning. J. Softw. **34**(1), 197–229 (2023)
5. Qolomany, B., Ahmad, K., Al-Fuqaha, A., et al.: Particle swarm optimized federated learning for industrial IoT and smart city services. In: Proceeding of 2020 IEEE Global Communications Conference, pp. 1–6. IEEE, Taipei, Taiwan (2020)
6. Zhu, H., Jin, Y.: Multi-objective evolutionary federated learning. IEEE Trans. Neural Netw. Learn. Syst. **31**(99), 1–13 (2019)
7. Zhong, J., Wu, Y., et al.: Multi-objective federated learning evolutionary algorithm based on improved NSGA-III. Comput. Sci. **50**(04), 333–342 (2023)
8. Zhiyong, H., Qiancheng, Y., et al.: Federated learning evolutionary algorithm based on multi-objective optimization. Appl. Res. Comput. **41**(02), 415–420 + 437(2024)

Unlocking Your Sales Insights: Advanced XGBoost Forecasting Models for Amazon Products

Meng Wang, Yuchen Liu, Gangmin Li, Terry R. Pyane, Yong Yue, and Ka Lok Man

Abstract One of the important factors of profitability is the volume of transactions. An accurate prediction of the future transaction volume becomes a pivotal factor in shaping corporate operations and decision-making processes. E-commerce has presented manufacturers with convenient sales channels to, with which the sales can increase dramatically. In this study, we introduce a solution that leverages the XGBoost model to tackle the challenge of predicting sales for consumer electronics products on the Amazon platform. Initially, our attempts to solely predict sales volume yielded unsatisfactory results. However, by replacing the sales volume data with sales range values, we achieved satisfactory accuracy with our model. Furthermore, our results indicate that XGBoost exhibits superior predictive performance compared to traditional models.

Keywords Ensemble learning · Consumer electronics · Sales forecasting · Amazon · XGBoost · CatBoost · GBDT · E-commerce first section

M. Wang · Y. Liu · Y. Yue · K. L. Man (✉)
Department of Computing, Xi'an Jiaotong-Liverpool University, Suzhou, China
e-mail: Ka.Man@xjtlu.edu.cn

M. Wang
e-mail: Meng.Wang22@student.xjtlu.edu.cn

Y. Liu
e-mail: Yuchen.Liu21@student.xjtlu.edu.cn

Y. Yue
e-mail: Yong.Yue@xjtlu.edu.cn

G. Li
College of Industry-Entrepreneurs (CIE), Xi'an Jiaotong-Liverpool University, Suzhou, China
e-mail: gangmin.li@xjtlu.edu.cn

T. R. Pyane
Department of Computer Science, University of Liverpool, Liverpool, England
e-mail: T.R.Payne@liverpool.ac.uk

J. S. Park et al. (eds.), *Advanced Multimedia and Ubiquitous Engineering*, Lecture Notes in Electrical Engineering 1475, https://doi.org/10.1007/978-981-95-1565-3_23

1 Introduction

With the growth of the Internet and e-commerce platforms, e-commerce has become a key sales channel for manufacturers. Accurate sales forecasts on e-commerce platforms can help dealers and manufacturers cut inventory costs, boost profits, and adjust pricing strategies in time [1, 2]. Such forecasts also assist upstream suppliers in anticipating the demand for finished goods, lessening the demand distortion on the supply chain [3]. Accurate sales forecasting is vital for marketing strategies, efficient business operations, and profitability.

Many studies have achieved excellent results for commodity sales forecasting using different technologies and models. Numerous scholars have made substantial progress in forecasting accuracy by refining existing time series models or adopting a multi-model integration approach [4, 5]. Conversely, the emergence of machine learning and deep learning techniques has showcased remarkable proficiency in addressing scenarios with limited historical data or subtle patterns [6, 7]. A rigorous study conducted by Dipillo and colleagues compared the predictive power of the support vector machine (SVM) against the ARIMA model using a standardised dataset for forecasting transaction values [8]. The results demonstrated SVM's potential in navigating the intricacies of sales forecasting, especially in scenarios where historical data is scarce or exhibits unpredictable fluctuations. Additionally, the numerous benefits of machine learning and deep learning techniques have sparked a significant amount of research examining the impact of product features on sales and developing methods for forecasting e-commerce revenue [9, 10]. The factors that impact sales, such as holiday periods [11, 12], contribute significantly to e-commerce sales forecasting. Hence, including multidimensional and multifaceted features is crucial in generating accurate predictions.

The sales data employed in this study is sourced from the Amazon website, particularly consumer electronics products. We aim to predict sales volumes by leveraging XGBoost algorithms with this dataset. Furthermore, these sales volumes are communicated as range estimates instead of specific values, a factor that potentially undermines the model's predictive precision. Considering these limitations, this study further subdivides the sales volume data based on the initial dataset to evaluate the model's practical reliability in short-term sales forecasting. The remainder of this paper is organised as follows: Sect. 2 details the methodology, including data sources, preprocessing, and model implementation. Section 3 presents the experimental results, focusing on XGBoost's predictions in consumer electronics and comparing them to other models. Lastly, Sect. 4 concludes by discussing limitations and outlining future research work.

2 Methodology

XGBoost is an additive model aggregating multiple decision tree predictive outcomes. Specifically, it comprises an ensemble of k-base learners. The overall model can be expressed as follows:

$$\hat{y}_i = \sum_{k=1}^{K} f_k(x_i), f_k \in \mathrm{F} \tag{1}$$

In formulation (1), the K represents the number of the trees; f_k is a specific tree; and $\hat{y}_i$ is the model output for a given sample length n and the number of features m, then:

$$D = \{(x_i, y_i)\}\big(|D| = n, x_i \in R^m, y_i \in R\big) \tag{2}$$

In Eq. (2), x_i represents the input of the i sample; y_i represents the output corresponding to the input of sample i; F is the space of the CART tree, which can be expressed as:

$$F = \big\{f(x) = w_q(x)\big\}\big(q : R^m \rightarrow T, w \in R^T\big) \tag{3}$$

where, q represents the structure of the tree; T is the child node of the tree; $f(x)$ is the CART tree structure q and child node weight w. By incorporating the regularisation term into the objective function:

$$Obj = \sum_i l(\hat{y}_i, y_i) + \sum_k \Omega(f_k) \tag{4}$$

$$\Omega(f) = rT + \frac{1}{2}\lambda\|w\|^2 \tag{5}$$

In Eq. (4), $\hat{y}_i$ and y_i represent the predicted value and actual label value of the i-th sample respectively; γ and λ are weight coefficients; T is the total number of leaves; w represents the model leaf node weight. Obj is the objective function, $\sum_i l(\hat{y}_i, y_i)$ is the loss error which represents the error between the predicted value and the true value and the second half is the regular term.

Table 1 Feature description

Num	Feature name	Feature meaning	Type
1	Products	Commodity classification	Object
2	Brand	Brand of commodity and object	Object
3	Colour	Color of commodity and object	Object
4	Manufacturer	Manufacturer of commodity	Object
5	Price	Price of commodity	Object
6	Rating	The customer's rating of the commodity	Float
7	Number of rating	The number of customer reviews	Float
8	Shipment	Shipment and time spent on logistics (day)	Float
9	Weight pounds	Weight pounds and the weight of commodity (pounds)	Float

3 Experiment

3.1 Data

The datasets employed in this study were derived from product information on the Amazon website, comprising 1565 distinct products. This comprehensive dataset encompasses various attributes such as product details, brand information, colour variations, manufacturer specifications, item weights, pricing details, customer ratings, review counts, sales volumes, and logistics timeframes. The primary product categories represented in the dataset include wireless headphones, gaming keyboards, computer mice, and air fryers. A detailed breakdown of the feature definitions is presented in Table 1.

3.2 Feature Engineering

This study employs several strategies tailored to different features to handle missing values. Firstly, for features where a missing value genuinely represents zero, such as ratings, review counts, and sales, all missing entries are replaced with zero to ensure practical reliability. Secondly, when encountering missing brand or manufacturer data (which often overlap), information from the other attributes is utilised to fill the gap. For missing logistics time and product weight values, the initial approach involves substituting them with the average values of similar products from the same brand and type. If such data is unavailable, the average values of products within the same type are used. Regarding colour features, specific shade descriptors (e.g., "dark grey") are simplified to their generic counterparts (e.g., "grey"). Products exhibiting multiple colours are categorised accordingly: dual-coloured products are labelled composite, while those with three or more colours are designated as "colourful". Furthermore, this study applies one-hot encoding to categorical variables such as categories, brands, colours, and manufacturers. The dataset is partitioned for prediction purposes into a training set comprising 80% of the data and a test set containing the remaining 20%.

3.3 Sales Prediction

The initial prediction evaluation in this article focuses on forecasting sales volume using the original dataset. While the outcomes were not optimal, with all five models exhibiting an MSE exceeding 200,000, it is noteworthy that XGBoost demonstrated superior performance compared to traditional machine learning models. The evaluation of the model's predictive capabilities posed a challenge due to the considerable magnitude of the MSE. Consequently, we undertook enhancements based on XGBoost to mitigate the excessive MSE stemming from the original data. However, it should be acknowledged that inaccuracies in the original dataset about product sales volume might have influenced these results. Hence, enhancing the quality of the original data is paramount for boosting model prediction proficiency.

Given the ambiguous data presented on the Amazon website, this study aims to alleviate bias in model predictions by categorising sales volume into specific ranges. Specifically, we have segmented the sales volume into eight groups: 0–50, 50–100, 100–300, 500–1000, 1000–3000, 3000–5000, and 5000–10,000. Concurrently, we substituted the original dataset's sales volume data with these newly generated range values and reintroduced them into the model for predictive purposes. The predicted outcomes of the range values are documented in Table 2.

The results documented in Table 2 reveal substantial enhancements in the predictive performance of each model when applied to range values as opposed to ordinary values. Regarding MSE, GBDT attained a value of 3.45, Bayes achieves 4.67, and SVM yields 3.51. However, XGBoost emerges as the superior model, exhibiting an MSE of only 1.93. This outstanding performance indicates that the average deviation between the predicted and actual values for XGBoost is less than two ranges. Furthermore, XGBoost showcases its dominance in prediction accuracy by attaining the lowest RMSE (1.389) and MAE (1.072) among all five models considered. These findings underscore the exceptional predictive capabilities of XGBoost, characterised by minimal average prediction errors. The linear regression exhibited

Table 2 Forecasting result for range

Model	MSE	RMSE	MAE
GBDT	3.45	1.86	1.23
XGBoost	1.93	1.39	1.07
Linear	1.06E+14	1.03E+07	2.29E+06
Bayes	4.47	2.11	1.66
SVM	3.52	1.88	1.51

relatively unusual results in this predictive experiment, with all three performance indicators exceeded 2,000,000. This observation suggests that the processing proficiency of linear regression in non-time series problems is considerably inferior to our enhanced XGBoost model. The RMSE metric further supports XGBoost's superiority over the other four models in addressing significant error problems. Given the relatively small size of our dataset and its vulnerability to overfitting and extreme values, the performance of the alternative models in terms of RMSE needs to improve compared to XGBoost's proficiency in this area. Similarly, this holds for MAE as well. In conclusion, XGBoost performs remarkably in predicting electronic consumer product sales, highlighting its utility and effectiveness.

4 Conclusion

The current research utilises a dataset of 1565 distinct product data entries spanning four unique categories from the Amazon website. This study incorporates various attributes of these products' non-time series to formulate a suitable dataset. Subsequently, this dataset is employed in an XGBoost model to forecast sales volume and range. A comparative analysis with traditional machine learning models assessed the model's predictive performance. The findings reveal that the XGBoost model demonstrates superior performance across three primary evaluation metrics: Mean Squared Error (MSE), Root Mean Squared Error (RMSE), and Mean Absolute Error (MAE). This underscores the model's reliability and effectiveness in predicting sales outcomes, particularly for consumer electronics products characterised by rapid product iteration and low short-term repurchase rates.

Despite achieving promising results with the methodology employed in this study, it is essential to acknowledge certain limitations. This study focused on specific commodity categories with potential substitutes, such as air fryers vs. conventional fryers. Future research should aim for a broader scope, particularly in electronics, to enhance the reliability of sales predictions. In future research directions, the primary objective will be to augment the dataset with additional product entries and expand the scope of product categories. Furthermore, integrating other machine learning models with XGBoost holds promise for developing a more reliable and robust predictive framework. Lastly, future studies aim to explore the correlation between features and prediction outcomes to construct a more efficient and accurate prediction model.

Acknowledgements This work is partially supported by the XJTLU AI University Research Centre and Jiangsu Province Engineering Research Centre of Data Science and Cognitive Computation at XJTLU. Also, it is partially funded by the Suzhou Municipal Key Laboratory for Intelligent Virtual Engineering (SZS2022004) as well as funding: XJTLU-REF-21-01-002 and XJTLU Key Program Special Fund (KSF-A-17).

References

1. Nguyen, H.D., Tran, K.P., Thomassey, S., Hamad, M.: Forecasting and anomaly detection approaches using LSTM and LSTM autoencoder techniques with the applications in supply chain management. Int. J. Inf. Manage. **57**, 102282 (2021)
2. Liu, Y., Man, K., Li, G., Payne, T.R., Yue, Y.: Dynamic pricing strategies on the internet. In: Proceedings of International Conference on Digital Contents: AICo (AI, IoT and Contents) Technology (2022)
3. Lee, H.L., Padmanabhan, V., Whang, S.: The bullwhip effect in supply chains (1997)
4. Nguyen, H.V., Naeem, M.A., Wichitaksorn, N., Pears, R.: A smart system for short-term price prediction using time series models. Comput. Electr. Eng. **76**, 339–352 (2019)
5. Liu, J., Liu, C., Zhang, L., Xu, Y.: Research on sales information prediction system of e-commerce enterprises based on time series model. IseB **18**, 823–836 (2020)
6. Liu, Y., Man, K.L., Li, G., Payne, T., Yue, Y.: Enhancing sparse data performance in e-commerce dynamic pricing with reinforcement learning and pre-trained learning. In: 2023 International Conference on Platform Technology and Service (PlatCon), pp. 39–42. IEEE, USA (2023)
7. Liu, Y., Mikriukov, D., Tjahyadi, O.C., Li, G., Payne, T.R., Yue, Y., Siddique, K., Man, K.L.: Revolutionising financial portfolio management: the non-stationary transformer's fusion of macroeconomic indicators and sentiment analysis in a deep reinforcement learning framework. Appl. Sci. **14**(1), 274 (2023)
8. Di Pillo, G., Latorre, V., Lucidi, S., Procacci, E.: An application of support vector machines to sales forecasting under promotions. 4OR **14**, 309–325 (2016)
9. Bandara, K., Shi, P., Bergmeir, C., Hewamalage, H., Tran, Q., Seaman, B.: Sales demand forecast in e-commerce using a long short-term memory neural network methodology, pp. 462–474. Springer, Berlin (2019)
10. Liu, Y., Man, K.L., Li, G., Payne, T., Yue, Y.: Evaluating and selecting deep reinforcement learning models for optimal dynamic pricing: a systematic comparison of PPO, DDPG, and SAC. (CCEAI 2024), pp. 215–219. ACM (2024)
11. Zhan, C., Wu, F., Huang, Z., Jiang, W., Zhang, Q.: Analysis of collective action propagation with multiple recurrences. Neural Comput. Appl. **32**, 13491–13504 (2020)
12. Wen, Q., Zhan, C., Gao, Y., Hu, X., Ngai, E., Hu, B.: Modeling human activity with seasonality bursty dynamics. IEEE Trans. Industr. Inf. **16**, 1130–1139 (2019)

Application Development of an Enhanced Autonomous Valet Parking System

Changhao Piao, Yi Xu, Jong Hyuk Park, and Mingjie Liu

Abstract At present, the urban traffic environment is very complex. In order to solve the social problem of shortage of parking spaces in the city and the tedious and troublesome task of finding parking spaces in the city, this thesis designs and develops an *Enhanced Autonomous Valet Parking (EAVP)* App based on 5G and V2X communication technologies. The App is equipped on the Android platform, and users can send a request to the EAVP cloud server through the App, and can view the information of the nearby parking lot and the vehicle itself in real time, grasp the vehicle status, and realize one-click autonomous parking and automatic pick-up.

Keywords Enhanced Autonomous Valet Parking (EAVP) · App · Android

1 Introduction

With the rapid advancement of urbanization and the continuous growth of the number of cars, the shortage of parking spaces has become a problem that cannot be ignored in the daily life of urban residents [1]. To address this challenge, this paper combines advanced technologies such as 5G and V2X communications [2] to design and develop an app called *Enhanced Autonomous Valet Parking (EAVP).* Through the app, users can obtain information about nearby parking lots and the status of their vehicles in real time, and realize the convenient operation of one-click autonomous parking and automatic pick-up. The introduction of this technology is expected to effectively alleviate the problem of urban parking, improve the intelligent level of

C. Piao (✉) · Y. Xu · M. Liu
Department of Automation, Chongqing University of Posts and Telecommunications, Chongqing, China
e-mail: piaoch@cqupt.edu.cn

M. Liu
e-mail: liumj@cqupt.edu.cn

J. H. Park
Seoul National University of Science and Technology, Seoul, South Korea

J. S. Park et al. (eds.), *Advanced Multimedia and Ubiquitous Engineering*, Lecture Notes in Electrical Engineering 1475, https://doi.org/10.1007/978-981-95-1565-3_24

urban transportation, and bring new convenience and comfort to the travel experience of urban residents.

2 EAVP App Function Design and Implementation

The *Enhanced Autonomous Valet Parking (EAVP)* app is designed to solve the city's parking challenges and make urban transportation smarter. The app is designed with functions such as one-click autonomous parking and pick-up, real-time access to parking lot information, vehicle status monitoring, message notification and early warning. In addition, the system needs to be designed to meet the following requirements:

(1) The App is required to have a good interaction mode [3], and the UI design should meet the needs of most users;
(2) Real-time detection provides accurate data to users, data transmission is flexible, and the robustness of the program is good;
(3) The system shall ensure the stability of operation, and take strict security measures and privacy protection mechanisms to ensure the safety of users' vehicles and personal information [4].

2.1 EAVP App System Architecture

The functional architecture of the EAVP App module is shown in Fig. 1, which is mainly divided into login registration and use. And the user interface is composed of three parts: homepage, message record, and me, of which the home page is composed of three main functional modules: automatic parking, real-time monitoring, and automatic car pick-up.

2.2 EAVP App Function Module Design

The core module of the EAVP App is the vehicle control and monitoring module, which is described below as the automatic parking and pick-up module, real-time monitoring module, message notification and early warning module.

2.2.1 Automatic Parking and Pick-Up Module

In As an application developed for the autonomous driving function of the vehicle, the module supports the function of one-click parking and pick-up [5]. When the user reaches the target parking lot within one kilometer, he logs in to his account with the

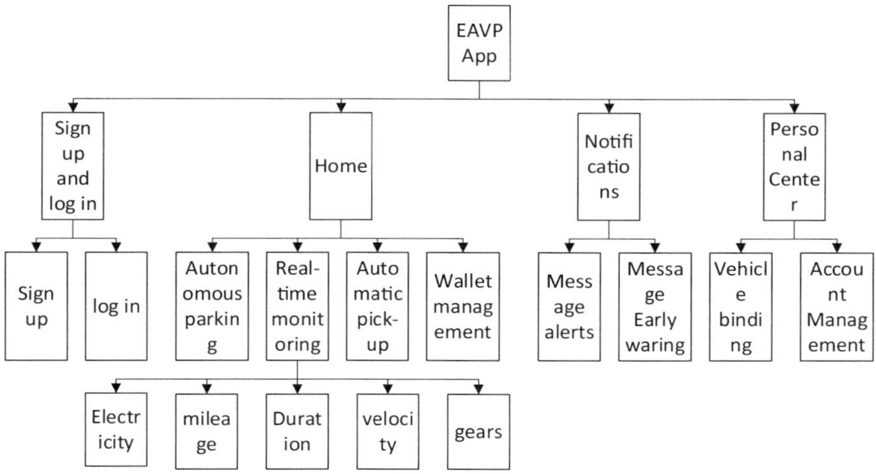

Fig. 1 EAVP app system framework diagram

App, and the App sends the current location information of the vehicle to the cloud server. The user clicks the *Automatic Parking* button on the home page of the app to enter the function module, confirms the path and parking lot information, starts parking, and the App sends the confirmation information to the cloud server, so as to start the vehicle to the nearest parking lot. When the user needs to pick up the car from the parking lot, the user only needs to click the *Automatic Pick-up* button on the home page of the App to enter the function, the App sends a request to download the map path data of the cloud server, the App receives the map data and displays it to the user after the server sends the map path data, and after the user selects the end point, click the "Confirm Pick-up" button and send a confirmation pick-up request. The key code snippets and effects are shown in Fig. 2.

2.2.2 Real-Time Monitoring Module

The user enters the APP and clicks the "Real-time Monitoring" button to enter the function module; After the APP sends a map request to the cloud, it can download the data transmitted by the cloud server and display the driving status, location information, and parking completion time of the vehicle on the specified path on the APP interface [6]. The key code snippets and effects are shown in Fig. 3.

2.2.3 Message Notification and Early Warning Module

In the upper part of the homepage, the user can view the current driving situation of the vehicle, and some important message reminders will also be displayed to the

```
//marker点击后的回调
override fun onMarkerClick(p0: Marker): Boolean {
    val latLng = p0.position
    endlat = latLng.latitude
    endlog = latLng.longitude
    //首先判断infowindows
    if (p0.isInfoWindowShown) {
        //infoWindow
        if (endlat != carlat && endlog != carlog) {
            AlertDialog.Builder(this).setTitle("确认提示")
                .setMessage("确认泊车到此停车场吗")
                .setPositiveButton("是", { dialogInterface: DialogInterface, i: Int ->
                    viewModel.postEparking(endlog.toString(), endlat.toString())
                    //将infoWindow隐藏
                    p0.hideInfoWindow()
                })
                .setNegativeButton("否", { dialogInterface: DialogInterface, i: Int ->
                    dialogInterface.dismiss()
                }).show()
        }
    }else{
        p0.showInfoWindow()
    }
    return true
}
```

(a)

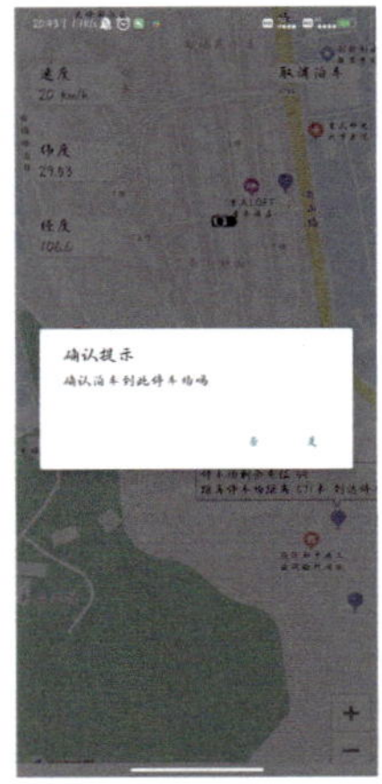

(b)

Fig. 2 **a** Code snippet diagram, **b** rendering of parking function

```
private fun initObserve() {
    //获取车辆的位置
    viewModel.postInitCarinf()
    //获取车辆的行驶状态
    viewModel.postCarNum()
    //监听车辆的位置 并显示在地图上
    viewModel.carinfInitBean.observe {
        carlat = it[0].carlat
        carlog = it[0].carlog
        infoLogLat(carlat.toFloat(),carlog.toFloat())
        carPosition()
    }
    viewModel.carinfaBean.observe {
        mTvSpeed.text = "速度      ${it[0].carspd} km/h"
    }
    //监听车辆的行驶状态
    viewModel.carNotification.observe {
        if (it == "泊车进行中") {
            isParking = true
            //如果是泊车进行中，就获取到储存的数据进行计算路径
   //       notCalculateRoute()
            viewModel.postCarinf()
            ......
```

(a)

(b)

Fig. 3 **a** Code snippet diagram, **b** rendering of real-time monitoring function

user in the form of a floating window. When it is difficult to complete automatic parking in the parking lot, non-vehicle faults, and software errors cause collisions and it is difficult to complete the automatic parking function, the APP should quickly receive the cloud server data and quickly notify the user for processing. The key code snippets and effects are shown in Fig. 4.

```
viewModel.carinfaBean.observe {
    mTvSpeed.text = "速度        ${it[0].carspd} km/h"
}
viewModel.parkingBean.observe {
    val arrayLatLng = ArrayList<LatLng>()
    for (i in it) {
        val new = LatLng(i.parklat, i.parklog)
        arrayLatLng.add(new)
    }
    for (i in 0 until  arrayLatLng.size) {
        //首先创建一个MarkerOptions的对象，用来显示标记点，后续应该通过数组来储存
        val markerOption = MarkerOptions()
            .title("停车场剩余车位 ${it[i].parkplace.toInt()}")
            .snippet("距离停车场距离 ${it[i].distance}米 到达停车场时间大约 ${it[i].duration / 60}分")
        //为标记点添加position，在哪个经纬度显示坐标
        markerOption.position(arrayLatLng[i])
        //将marker添加到aMap中
        val marker =  aMap.addMarker(markerOption)
    }
}
```

(a)

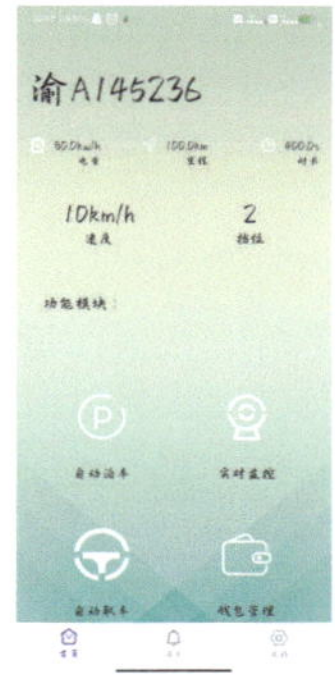

(b)

Fig. 4 **a** Code snippet diagram, **b** rendering of message viewing function

3 EAVP App Interface Design and Implementation

When users use the app for the first time, they need to register and log in before they can use it, and enter the account, account password and login button on the login page. The registration interface requires the user to use an email address, phone number and their own password as a registration condition, and click the "Register" button to send data to the server and jump to the login interface after the registration is completed. After the login is completed, the user enters the interface prepared for the first time, and the interface can collect the user's vehicle information (i.e., the vehicle VIN code), and after the information entry is completed, if the vehicle cannot use the AVP function, a pop-up window prompts the user that the user cannot bind; After the AVP can be used to bind the vehicle, the user enters the user interface. The user interface consists of three parts: Homepage, Message Records, and Me; The functional layer of the EAVP App is a collection of various data and instructions provided to the user, allowing the user to understand and operate the App. The background processing layer performs various logical processing on the instructions sent by the user interaction, calls the interface to transmit to the network layer, and the network layer returns the data resources, and the model layer temporarily stores the resource information to prevent errors in the data resources, and all the interaction data is exchanged in the MySQL database. Figure 5 shows the flow chart of the App system.

In terms of interface, in order to ensure that the interface response speed is fast enough and cross-platform, the data format for transmission is JSON format. At the same time, the APP interface needs to ensure on-demand allocation (that is, what data is returned if it is needed) to ensure the data processing efficiency of the APP. The interface should also be extensible. Table 1 lists the data to be transferred.

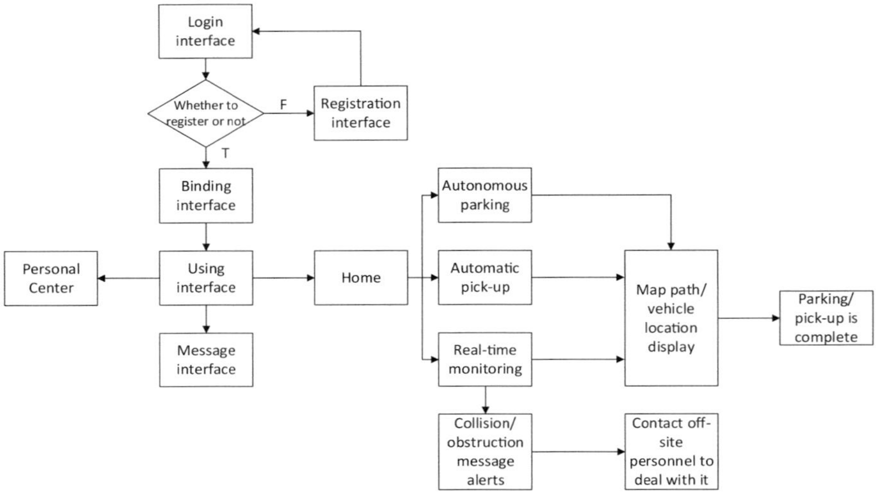

Fig. 5 EAVP app usage flowchart

Table 1 EAVP App interface design and transfer data table

Interface	Transfer of data (JSON)
Registration interface	Email, phone, username and password set by the user
Login interface	Accounts and Passwords
Personal center	Personal and vehicle information (i.e. VIN code, license plate number)
Message interface	Message records, prompts, and warning information records during use
Home interface	Download Data Request, Parking Confirmation Request, Pickup Confirmation Request, Map Data, Route Data, Vehicle Control Request, Vehicle Control Data, Real-Time Vehicle Status Data

4 Conclusion

Through the research and design of this paper, we have successfully developed an enhanced Automated Valet Parking (EAVP) App based on advanced technologies such as 5G and V2X communications. In the design and development process, we fully consider the needs of users and use scenarios, through good interaction and humanized interface design, so that users can easily get started and enjoy convenient parking services. At the same time, we also pay attention to the stability and security of the system, and have adopted strict security measures and privacy protection mechanisms to ensure the safety of users' vehicles and personal information. Through the EAVP App designed and developed in this paper, we are expected to effectively alleviate the problem of urban parking difficulties, improve the intelligent level of urban transportation, and bring new convenience and comfort to the travel experience.

Acknowledgements This research was funded by the National Key Research and Development Program of China, grant number No. 2022YFE0101000.

References

1. Ćelić, J., Mandžuka, B., Tomas, V., et al.: Driver-centric urban route planning: smart search for parking. Sustainability **16**(2) (2024)
2. Filippo, M., Nicola, C., Manuela, M.: Improving urban viability through smart parking. International Journal of Parallel, Emergent and Distributed Systems **38**(6), 522–540 (2023)
3. Kumar, K., Singh, V., Raja, L., et al.: A review of parking slot types and their detection techniques for smart cities. Smart Cities **6**(5), 2639–2660 (2023)
4. Ali, J., Khan, F.M.A.: Trust-based secure parking allocation for IoT-enabled sustainable smart cities. Sustainability **15**(8) (2023)
5. Elfaki, O.A., Messoudi, W., Bushnag, A., et al.: A smart real-time parking control and monitoring system. Sensors **23**(24) (2023)
6. Zhi, L.Q., Kum, T.D.T, Swee, P.K.: Integrated smart public parking system for Malaysia. J. Phys.: Conf. Ser. **2523**(1) (2023)

Research on Video Stream Continuity Status Prediction of Connected Vehicles in V2X Mode

Changhao Piao, Dong Yin Wang, Jungho Kang, and Ping Liu

Abstract A V2V real-time video stream overdata transmission optimization algorithm is designed to improve the quality of the receiver's accident video playback. The method predicts the current transmission state based on kalman filter, and adds RTSS to make further accuracy of the prediction results. Mathematical analysis and simulation data show that this method can make the receiver get a better playback experience and improve the reliability and real-time transmission of video stream.

Keywords V2V · Kalman forecast · RTSS · Video transmission

1 Introduction

In order to ensure traffic safety and improve traffic efficiency, the intelligent Transportation System (ITS) has developed rapidly around the world. As an intelligent system to improve traffic safety and efficiency and reduce traffic accidents [1], the Internet of Vehicles (IoVs) supports information interaction between vehicles and can provide vehicle users with real-time traffic information, accident warning, route planning, infotainment and other applications and services, which has great potential in improving traffic safety and efficiency and improving the comfort of drivers and passengers, etc., and has attracted great attention from researchers [2].

Based on in-depth analysis and research on the efficiency characteristics and network structure of V2V communication channels, this paper proposes a V2V-based video stream transmission state prediction algorithm to improve the playback quality

C. Piao · D. Y. Wang · P. Liu (✉)
Department of Automation, Chongqing University of Posts and Telecommunications, Chongqing, China
e-mail: liuping_cqupt@cqupt.edu.cn

C. Piao
e-mail: piaoch@cqupt.edu.cn

J. Kang
Department of Software Engineering, Baewha Women's University, Seoul, South Korea
e-mail: kjh7548@naver.com

J. S. Park et al. (eds.), *Advanced Multimedia and Ubiquitous Engineering*, Lecture Notes in Electrical Engineering 1475, https://doi.org/10.1007/978-981-95-1565-3_25

of the video stream information received by the rear vehicle. The second algorithm predicts the number of packets lost and sent in the current channel transmission based on Kalman filter. The adaptive threshold algorithm is combined with the output of the current network state to dynamically allocate the transmission performance, and improve the reliability and real-time transmission of video stream.

2 Related Work

In V2V communication, it is inevitable that vehicles will not receive information in time and lose information. Literature [3] introduced the effect of queue overflow packet loss in channel transmission to minimize the long-term average packet loss rate of total message set. Literature [4] introduced an improved cross-layer rate optimization model for message set transmission over wireless fading channels to find the optimal transmission rate strategy and obtain a new adaptive control optimization algorithm for data transmission rate. Literature [5] used SAE congestion control in LTE-V2X to alleviate the problem of continuous packet loss through wider bandwidth and message set retransmission. Literature [6] introduced mobile edge computing technology and proposes an efficient point-to-point information transmission mechanism for vehicle networking, which simultaneously satisfies the information transmission between vehicles and vehicles and between vehicles and roads, reduces the service response delay and improves the information transmission efficiency of vehicle networking.

3 V2V-Based Prediction of Video Stream Transmission Status

3.1 Network State Prediction Based on Kalman Filter

First, in V2V communication, the number of packets sent and the packet loss book are taken as predictors to obtain the packet loss rate. The state vector and state transition matrix for kalman filter are defined as follows

$$s(t) = \begin{bmatrix} q(t) \\ q'(t) \\ q''(t) \end{bmatrix} \tag{1}$$

Equation (1) is the state vector used to predict the number of packet sending, where $q(t)$ is the number of video stream packets sent by the head car, $q'(t)$ represents the first-order variation of $q(t)$ over time and the packet sending speed, and $q''(t)$

represents the second-order variation of $q(t)$ over time, that is, the packet sending acceleration.

Because the number of packets sent and packet books lost in V2V video streaming communication are not linear. The state vector contains second-order variations. Compared with the state vector that only considers first-order variations, the packet loss rate can be predicted more accurately. The state transition matrix is as follows:

$$\Phi(\Delta t) = \begin{bmatrix} 1 & \Delta t & \frac{\Delta t^2}{2} \\ 0 & 1 & \Delta t \\ 0 & 0 & 1 \end{bmatrix} \tag{2}$$

In the equation, Δt is the transmission delay of each frame.

The state of the V2V communication network changes over time according to the state transition matrix [7]. Therefore, in the t interval, according to the number of packets lost, the packet lossing speed and acceleration at the time of $t - \Delta t$ can be calculated. According to the packet dropping speed and acceleration at the time of $t - \Delta t$, the change of packet dropping state at the time of t can be solved. For the multistate transition matrix represented by the state vector, see Eq. (3) and equation:

$$\begin{aligned} q(t) &= q(t - \Delta t) + q'(t - \Delta t) \cdot \Delta t + q''(t - \Delta t) \cdot \frac{\Delta t^2}{2} \\ q'(t) &= q'(t - \Delta t) + q''(t - \Delta t) * \Delta t \\ q''(t) &= q''(t - \Delta t) \end{aligned} \tag{3}$$

We design measurement vectors and observation matrices to design state measurement models. The measurement vector represents a matrix of possible measurements between the network state vectors. The number of packets sent and the number of packets lost are the white variables of the measurement vector. Therefore, the measurement vector can be expressed as Eqs. (4) and (5):

$$z(t) = (q) \tag{4}$$

$$zL(t) = (p) \tag{5}$$

The observation matrix is used to solve the measurement vector from the state vector, so according to the definition of the state vector and the measurement vector, it can be expressed as Eq. (6)

$$H(t) = (1, 0, 0) \tag{6}$$

Step 1. Initialization phase: The proposed Kalman filtering method describes the V2V packet loss state from the point of view of signal and measurement model.

$$X(t) = \Phi \cdot X(t-1) + w(t-1) \tag{7}$$

$$z(t) = H \cdot X(t) + n(t) \tag{8}$$

Step 2. prediction stage: prediction stage: Find a priori estimate $\tilde{X}^{-}(t)$ by $\tilde{X}^{-}(t) = \Phi\tilde{X}^{+}(t-1)$, and then we can derive the prediction error covariance $p^{-}(t)$ from $p^{-}(t) = \Phi p^{+}(t-1)\Phi^{T} + Q$.

Step 3. measurement/filtering stage: In the measurement stage, After acquiring this measurement value (z), the Kalman gain (K) can be computed by

$$K(t) = \frac{p^{-}(t) \cdot H^{T}}{H \cdot p^{-}(t) \cdot H^{T} + R} \tag{9}$$

Through the filtering process, we can find the posterioribandwidth estimate $\tilde{X}^{+}$ by

$$\tilde{X}^{+} = (1-H)\tilde{X}^{-} + z(t)K(t) \tag{10}$$

and the filtering error covariance $p^{+}(t)$ can be calculated by $p^{+}(t) = (1-K(t))p^{-}(t)$.

Step 4. Increase the time index t, and go back step 2 for the next time-slot.

3.2 RTSS Smooth

The Rauch-Tung-Striebel Smoother (RTSS) is a smoother that estimates the state at each moment based on getting all the estimates. After completing one KF transmission status estimation, one RTSS is connected, and the transmission statuzs state information obtained by KF estimation is reversely used to perform one smoothing operation to further improve the accuracy of transmission status estimation. The specific smoothing equation is as follows:

$$z_{i+1|i}^{(s)} = T_i z_i u_i \tag{11}$$

$$P_{i+1|i}^{(s)} = T_i P_i T_i^{\mathrm{T}} + Q_u \tag{12}$$

$$\boldsymbol{J}_i = \boldsymbol{P}_i \hat{\boldsymbol{A}}_i^{\mathrm{T}} (\boldsymbol{P}_{i+1|i}^{(s)})^{-1} \tag{13}$$

$$\boldsymbol{z}_i^{(s)} = \boldsymbol{x}_i + \boldsymbol{J}_i\left(\boldsymbol{z}_{i+1}^{(s)} - \boldsymbol{z}_{i+1|i}^{(s)}\right) \tag{14}$$

$$P_i^{(s)} = P_i + J_i\left(P_{i+1}^{(s)} - P_{i+1|i}^{(s)}\right)J_i^{\mathrm{T}} \tag{15}$$

In the formula, $z_{i+1|i}^{(s)}$ represents the prior state variable of RTSS, $P_{i+1|i}^{(s)}$ represents the prior covariance matrix of RTSS, $\boldsymbol{J}_i$ represents the gain of RTSS, z_i Represents the posterior state variable of RTSS, $P_i^{(s)}$ represents the posterior covariance matrix of RTSS, z_i is the posterior state variable estimated by RTSS, P_i Represents the posterior covariance matrix obtained by RTSS. We can expect that the relay smoothing operation can further improve the accuracy of channel estimation, but at the cost of computational complexity [8].

3.3 Adaptive Threshold Design

Due to the uncertainty of communication conditions caused by high-speed vehicle movement, the identification of current communication conditions requires a certain degree of adaptability. We designed an adaptive threshold to deal with such scenarios. The goal of the adaptive threshold block is to adapt the sensitivity of the algorithm to the packet loss status based on KF-RTSS.

$$Y_t = Y_{t-1} + \Delta t \cdot (X - Y_{t-1})K_u \tag{16}$$

In the formula, X is the packet loss rate of the KF-RTSS settlement, Y_t is the adaptive packet loss rate threshold, Δt is the time interval between two video stream packets, K_u is the gain. If the packet loss rate is greater than the adaptive threshold, frame complement is performed to optimize packet loss.

4 Simulation Results

Further, it is verified by simulation that the algorithm can predict packet loss through the previous frame of the image frame, so as to ensure the integrity of the video. The packet loss of the optimized video data is shown as follows. The video footage was collected in real time by real vehicles at the relative distance of 25 m, 50 m and 100 m, and the collected data were analyzed corresponding to Figs. 1, 2 and 3 respectively.

As can be seen from Figs. 1, 2 and 3, the blue line represents the frame loss of the original transmission of V2V, and the red line represents the frame loss after KF-RTSS algorithm. If frame loss occurs, the frame number is reduced to 0, otherwise it will be a continuity label. It is not difficult to see that our algorithm has a significant improvement on frame loss in V2V video stream transmission.

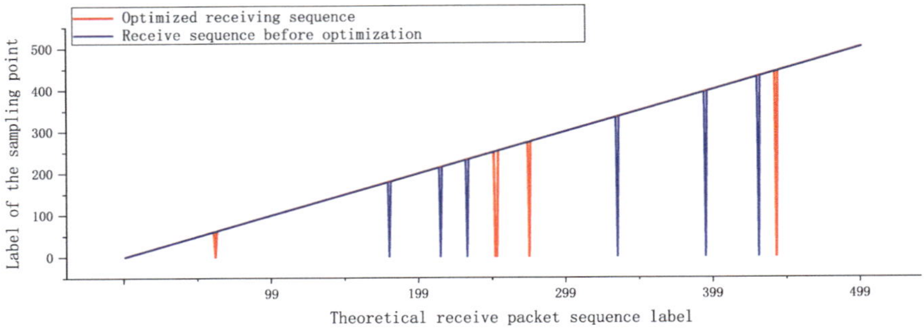

Fig. 1 25 m Video data after frame optimization

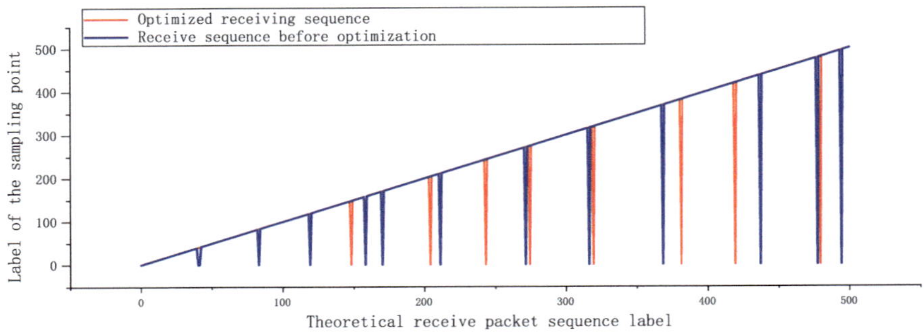

Fig. 2 50 m Video data after frame optimization

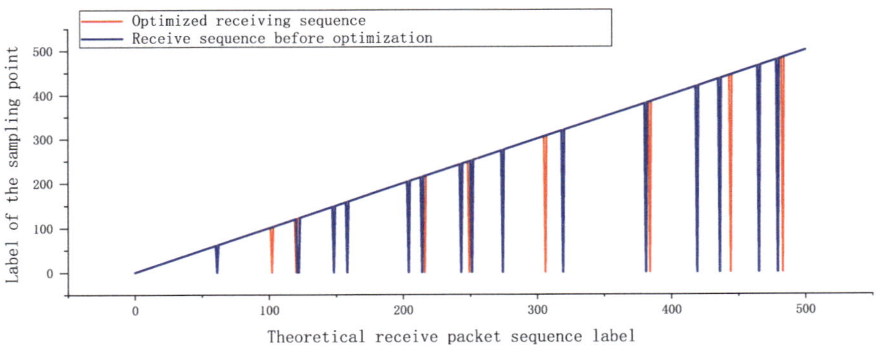

Fig. 3 100 m Video data after frame optimization

5 Conclusion

This study proposes an optimized V2V over the line video transmission frame loss prediction algorithm, which not only ensures the reliability of data when the rear vehicle receives video information from the front vehicle, but also meets the real-time requirements of vehicle networking. The measurement data is used to estimate the future packet loss rate through KF-RTSS, and timely perform frame repair for the next packet loss time to improve vehicle communication efficiency.

Acknowledgements This research was funded by the National Key Research and Development Program of China, grant number No. 2022YFE0101000.

References

1. Benalia, E., Bitam, S., Mellouk, A.: Data dissemination for Internet of vehicle based on 5G communications: a survey. Trans. Emerg. Telecommun. Technol. **31**(5), 1–34 (2020)
2. Soleimani, P., Marvasti, M.R.B., Ghorbanzadeh, P.: A hybrid traffic management method based on combination of IoV and VANET network in urban routing for emergency vehicles. In: 2020 4th International Conference on Smart City, Internet of Things and Applications (SCIOT). IEEE, pp. 58–65 (2020)
3. Yu, R.: Research on Collaborative Communication Technology of Highly reliable Relay Network in Vehicle networking. Donghua University (2023). 10.27012
4. Zhengyong, F.: Improvement of cross-layer optimization model for fading channel packet transmission. Comput. Eng. **42**(11), 125–130 (2016)
5. Li, Y., Wang, N., Lv, W.: Research on resource allocation algorithm of V2V communication based on service heterogeneity in cellular vehicle networking. J. Electron. Inf. **45**(01), 235–242 (2023)
6. Li, L., Wu, X., Zhu, P.: A safe transmission scheme of road condition information in vehicle-connected environment. Firepower Command. Control.
7. Duan, Y.: Analysis of Event-triggered Predictive Control for stochastic Systems with time Delay. Nanjing University of Posts and Telecommunications (2023)
8. Yong, L., Ying, C.: Highly reliable and robust V2V channel estimation of UKF-RTSS based on base extension model. J. Electron. Inf. Technol. **44**(05), 1792–1799 (2022)

Differential Attention Based Dual-Branch Vehicle Re-identification Network

Chenchen Zhang, Gen Zhao, Jing Wang, and Xu Zhang

Abstract In the field of intelligent transportation, vehicle re-identification (Re-ID) aims to accurately search for target vehicles from a large-scale vehicle gallery. Vehicle images tend to have more similar appearance and smaller differences among them. To address the challenges of vehicle re-recognition, we propose a differential attention based dual-branch vehicle re-identification network. Specifically, to suppress the environmental noise interference, we propose a spatial and channel attention mechanism of differential attention. Differential attention consists of difference spatial attention (DSA) and difference channel attention (DCA). We use the difference between a pair of feature maps obtained by double branching as a noise suppression factor. The experimental results show that the mAP and rank-k indicators of the model on VeRi-776 and the VehicleID are better than the results of the existing vehicle re-identification algorithms, which verifies the algorithm's effectiveness.

Keywords Vehicle re-id · Channel attention · Spatial attention · Dual-branch net

1 Introduction

With the increasing number of vehicles, intelligent transportation has also been developed and applied, therefore the intelligent demand corresponding to the monitoring system is getting higher and higher, and vehicle re-identification is getting more and more emphasis. The varying conditions under which the vehicle images were

C. Zhang · X. Zhang (✉)
Department of Computer Science and Technology, Chongqing University of Posts and Telecommunications, Chongqing, China
e-mail: zhangx@cqupt.edu.cn

G. Zhao
Chongqing Planning Exhibition Gallery, Chongqing, China

J. Wang
Changjiang Chongqing Communication Administration, Chongqing, China

J. S. Park et al. (eds.), *Advanced Multimedia and Ubiquitous Engineering*, Lecture Notes in Electrical Engineering 1475, https://doi.org/10.1007/978-981-95-1565-3_26

captured resulted in considerable variation within the instances, posing a significant challenge to vehicle re-identification.

Vehicle re-identification studies typically use CNN architectures as feature extractors, which are capable of learning both global and local features. Nevertheless, feature extraction alone can no longer satisfy the demands of today's complex and dynamic traffic scenarios. Recent work has proposed the following solutions in dealing with the above challenges. The representative approaches are categorized into the following three groups: (1) Methods based on various additional a priori knowledge [1, 2]. (2) A multi-branch structure is employed to enhance global discriminative features [4]. (3) Attention-based methods [3, 5, 6]. The attention mechanism enables the model to dynamically allocate focus to relevant regions within an image, capturing complex details and contextual cues that are essential for distinguishing individuals with similar appearances. Previous attention mechanisms have often focused on the characteristics of a single feature map, often neglecting the complementary features of the overall network.

In this paper, we introduce a Dual-branch Attention Differential (DAD) vehicle re-identification network that leverages attention mechanisms to capture spatial and channel relationships among local features. Our approach aims to mitigate the interference of background noise and redundant channel features around vehicles. Unlike traditional methods that overlook the interaction between the two branches, our approach harmonizes the outputs of the dual branches through the introduction of Differential Spatial Attention (DSA) and Differential Channel Attention (DCA) blocks, utilizing their differences as suppression factors. By optimizing the parameters of these components, our DAD network is able to dynamically adapt to different scenarios and handle challenges such as variations in viewpoint, occlusions, and background clutter.

The main contributions of this paper are as follows:

- We propose a novel dual-branch network based on differential attention (DAD), which combines spatial and channel attention mechanisms.
- We have designed a Dual-branch Differential Attention (DDA) module that integrates DSA block and DCA block. This module leverages the spatial-channel attention to optimize feature learning by utilizing the difference between the outputs of the dual branches as a noise suppression factor.

2 Proposed Methods

2.1 Overview

The overall architecture of the model is shown in Fig. 1. The initial three shared layers of the backbone network process the input image, extracting fundamental vehicle features. The output from the third layer undergoes a fourth layer where weight sharing is not employed. The difference between the two output features, along with

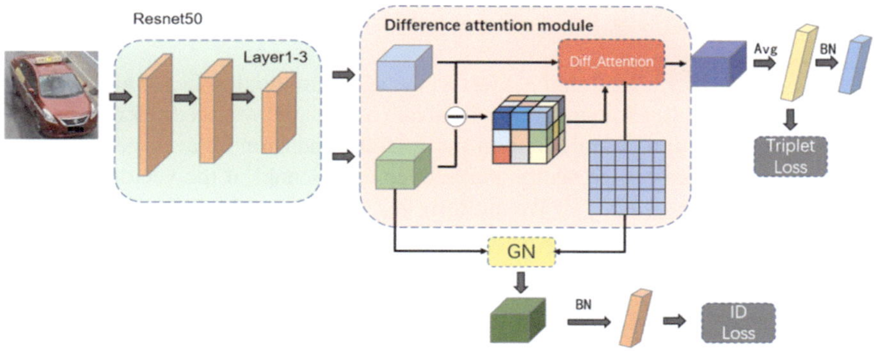

Fig. 1 The overall architecture of the proposed DAD model

the feature values from the first branch, serve as the input to the Differential Attention Module (DAM). The DAM's output, after passing through a pooling layer, is constrained by a triplet loss function. The second branch of the dual-branch architecture explores the dependencies between the DAM's output and its own features through a graph network. The final output is projected onto a linear layer, and the cross-entropy loss serves as the classification loss function.

2.2 *Differential Attention Module*

As previously mentioned, the Differential Attention Module (DAM) takes as input the outputs of two parallel branches, along with their corresponding difference. This difference is obtained through a subtraction operation performed between the two branches, specifically by subtracting corresponding elements from the feature maps to generate a new one. The resulting feature map, obtained through this subtraction, is then employed as a constraint within the subsequent spatial and channel attention mechanisms.

2.2.1 Differential Spatial Attention (DSA)

The Differential Spatial Attention block aims to generate spatial attention maps, constrained by the differential feature map, by considering relationships between a global spatial reference node and individual nodes. Let $X \in R^{C \times H \times W}$ be the input tensor of the DSA block, where C is the number of channels and H and W correspond to the height and width of the tensor respectively.

We pass the input X through three learnable weight matrices W_K, W_Q, and W_V, and reduce their channel dimensions from C to C/r. These operations can be defined as follows:

$$X_{K,Q,V} = W_{K,Q,V}(X) = \text{ReLU}(\text{BN}(\text{Conv}_{1*1}(X))). \tag{1}$$

where $X_{K,Q,V} \in R^{C/r \times H \times W}$ and r refers to the spatial reduction factor and BN here means batch normalization operation. To obtain a spatially averaged reference value, we compress the input feature map in the spatial dimension by average pooling to obtain an average reference vector, and reduce the channel of the vector from C to C/r similarly. So $V_A \in R^{C/r \times 1}$. We multiply X_V with V_A to obtain the spatial vector of correlations between features $V_S \in R^{1 \times HW}$.

To obtain a differential feature map, we differ the output features of the dual branch. In DSA, we take the average of the difference feature map, compress it in channel dimension to get the differential space feature map D_S, and dot-multiply it with Vs after reshape to get the final differential space feature map D'_S. These operations can be defined as follows:

$$D'_S = Vs \odot D_S = \left(X_V \otimes (V_A)^T\right) \odot \text{mean}(D_S). \tag{2}$$

where $\odot$ is broadcast element-wise multiplication and mean indicates the average value of all channels at each pixel. In order to obtain the spatial relationship between the nodes of the feature graph, we transpose X_K and multiply it with X_Q, and then reshape the resulting matrix of HW * HW in both the length and width dimensions to obtain $R_S \in R^{H \times W \times 2HW}$. Considering these three kinds of information are not in the same feature domain, we embed them respectively and concatenate them to get the spatial differential feature F_S:

$$F_S = \left[\text{reshape}(V_S),\ D'_S,\ R_S\right]. \tag{3}$$

We obtain feature F_S with rich structural information from which valuable knowledge can be mined to infer attention through learnable models. We get the final differential space attention feature graph by node integration, and to avoid gradient vanishing connect the initial features as residual blocks:

$$F'_S = X \odot \varphi_S(F_S) \tag{4}$$

where φ_S is implemented by a spatial 1×1 convolutional layer followed by BN and ReLU activation.

2.2.2 Differential Channel Attention (DCA)

Getting the output F'_S of the DSA, we designed a Differential Channel Attention block for learning C-dimensional differential attention feature maps.

Similar to spatial attention, we got $X_{K,Q,V} \in R^{C/r \times H \times W}$. To obtain an average reference value for each spatial point on the channel, we compute the average value on the channel. The result is multiplied with the VA to obtain the channel vector of correlations between features $V_C \in R^{1 \times C}$.

In DCA, we send the differential feature map through a global average pooling layer to obtain the differential channel feature map D_C and dot-multiply it with V_C to obtain the final differential channel feature map D'_C. These operations can be defined as follows:

$$D'_C = V_C \odot D_C = (X_V \otimes (V_A)^T) \odot \psi(D_S) \quad (5)$$

where ψ denotes average pooling operation. In order to obtain the channel relationship between the nodes of the feature graph, the transpose of X_K is multiplied with X_Q in the spatial dimension to obtain the spatial relationship matrix of $C \times C$. After reshaping operation and projection operation to get $R_C \in R^{1 \times c}$. we join the three obtained information to obtain the channel differential feature $F_C \in R^{3 \times c}$. F_C is passed through a 1×1 convolutional layer of one channel, followed by BN and ReLU activation, and residuals are connected to get the final F'_C.

3 Experiments

3.1 Implementation Details

Before training, we randomly applied crop, flip, and pad operations on the images with a certain probability. The images were then uniformly resized to 256 × 256 pixels. The models are trained on each dataset during 120 epochs with an Adam optimizer using a learning rate of 1e-4, a linear warm-up of 10 epochs and decrement steps at epochs 30, 60, and 90 with a factor of 0.1. The model was implemented using the PyTorch framework and trained and tested on an NVIDIA A100.

3.2 Experimental Results and Analysis

We evaluated our model on the VeRi-776 datasets and VehicleID datasets and compared its performance against various state-of-the-art methods. Specifically, our method achieved a higher mAP score, surpassing the baseline by 5.3% in the VeRi-776 datasets. Moreover, the Rank-1 and Rank-5 scores showed improvements of 2.6% and 1.2%, respectively, over the baseline. In the VehicleID datasets, our model has demonstrated strong performance in both Rank-1 and Rank-5 scores, as shown in Table 1).

Table 1 Comparison with state-of-the-art results (%) on VeRi-776 and Vehicle-ID

Method	Veri-776			Vehicle-ID					
	mAP	Rank-1	Rank-5	Small(800)		Medium(1600)		Large(2400)	
				Rank-1	Rank-5	Rank-1	Rank-5	Rank-1	Rank-5
PVEN [2]	79.5	95.6	98.4	84.70	97.00	80.60	94.50	77.80	92.00
SAVER [3]	79.6	96.4	98.6	79.90	95.20	77.60	91.10	75.30	88.30
TBE-Net [4]	79.5	96.0	98.5	–	–	–	–	–	–
SOFCT [5]	80.7	96.6	98.8	84.50	96.80	80.90	95.20	78.70	93.70
GRA-Net [6]	80.5	95.2	99.0	83.3	96.2	78.2	93.5	75.1	90.8
OURS	81.9	98.3	99.2	85.5	97.00	80.12	95.30	79.08	93.90

4 Conclusion

In this paper, we propose a novel dual-branch network based on differential attention, which combines a spatial channel attention mechanism with a new constrained feature map. The method effectively removes background noise and does not require additional annotations. We design a DAM module to implement this function, which adopts the differential attention mechanism of spatial and channel, and fuses the suppression factor derived from the difference of two branches to remove the background noise. Extensive experiments on the VeRi-776 and VehicleID datasets demonstrate the effectiveness of our proposed method. Visual comparisons with the baseline model show that our approach is more focused on the vehicles themselves, leaving room for improvement in the design of the inter-class distance computation. These results highlight the potential of our method for vehicle re-identification tasks.

Acknowledgements This research is supported in part by the Project of Key Laboratory of Tourism Multisource Data Perception and Decision, Ministry of Culture and Tourism, China (No. H2023009), and in part by the Open Fund of Key Laboratory of Monitoring, Evaluation and Early Warning of Territorial Spatial Planning Implementation, Ministry of Natural Resources (No. LMEE-KF2023003)

References

1. Huang, W., Zhong, X., Jia, X., et al.: Vehicle re-identification with spatio-temporal model leveraging by pose view embedding. Electronics **11**(9), 1354 (2022)
2. Meng, D., Li, L., Liu, X., et al.: Parsing-based view-aware embedding network for vehicle re-identification. In: Proceedings of the IEEE/CVF conference on computer vision and pattern recognition, pp. 7103–7112 (2020)

3. Khorramshahi, P., Peri, N., Chen, J., et al.: The devil is in the details: self-supervised attention for vehicle re-identification. In: Computer Vision–ECCV 2020: 16th European Conference, Glasgow, UK, Proceedings, Part XIV 16. Springer International Publishing, pp. 369–386 (2020)
4. Sun, W., Dai, G., Zhang, X., et al.: TBE-Net: a three-branch embedding network with part-aware ability and feature complementary learning for vehicle re-identification. IEEE Trans. Intell. Transp. Syst. **23**(9), 14557–14569 (2021)
5. Yu, Z., Huang, Z., Pei, J., et al.: Semantic-oriented feature coupling transformer for vehicle re-identification in intelligent transportation system. IEEE Trans. Intell. Transp. Syst. (2023)
6. Jiang, G., Pang, X., Tian, X., et al.: Global reference attention network for vehicle re-identification. Appl. Intell. 11328–11343 (2023)

A Blockchain-Based Archive Version Traceability and Management Method

Weiran Zhang, Jianrong Chen, Ming Huang, Wenhao Yang, and Xu Zhang

Abstract Archive provenance is a crucial aspect of ensuring the authenticity and security of information. In this paper, we introduces a blockchain-based electronic archive version traceability and management method. The method involves establishing a blockchain system architecture that includes archive information owner (AIO), visitor (AIV), and authorizing agency (AA) of electronic archives, enabling the desensitized encryption of archive information and the direct encryption of structured data. The storage of electronic archive information is completed through blockchain operations, ensuring transparency and traceability. The system is implemented with interplanetary file system (IPFS) and perform well in Chongqing municipal planning and natural resources archives.

Keywords Blockchain · Archives management · Version management

1 Introduction

In recent years, the exponential increase in digital data has been driven by the rapid development of information technology. Traditional methods of archive management often rely on centralized authorities and may be vulnerable to tampering, loss, or unauthorized access. Therefore, it is a critical challenge considering the authenticity, security and traceability [1, 2]. Existing methods proposed to manage archives in a centralized environment or general distributed file systems, e.g. HDFS. However, damaged nodes or attacks may lead the electronic documents to unreadable and unusable, and may even result in data loss, thereby reducing the efficiency of archive utilization. Additionally, poor management can lead to theft or tampering of information. Sharing archives between different organizations can be challenging, resulting

W. Zhang · J. Chen · M. Huang
Chongqing Municipal Planning and Natural Resources Archives, Chongqing, China

W. Yang (✉) · X. Zhang
Department of Computer Science and Technology, Chongqing University of Posts and Telecommunications, Chongqing, China
e-mail: 3471949741@qq.com

J. S. Park et al. (eds.), *Advanced Multimedia and Ubiquitous Engineering*, Lecture Notes in Electrical Engineering 1475, https://doi.org/10.1007/978-981-95-1565-3_27

in low efficiency in retrieval and utilization. Furthermore, tracing abnormal behaviors is a critical issue. To address these challenges, we propose a novel blockchain-based archive version traceability and management method.

In this paper, we introduce a method for tracing and managing archive document versions based on blockchain technology. Blockchain interoperability is essential for breaking inter-chain barriers and building a blockchain network [3]. The proposed approach is designed to establish a robust mechanism ensuring the security of electronic document traceability. Leveraging blind signature techniques, it ensures the anonymity of document uploaders while facilitating precise accountability during traceability procedures. Moreover, the method enhances traceability efficiency through the utilization of threshold-enhanced attribute-based encryption technology.

In summary, the proposed blockchain-based archive version traceability and management method offers a novel approach to ensuring the integrity, authenticity, and traceability of digital archives. By leveraging the inherent properties of blockchain and smart contracts, we aim to create a secure and transparent system for managing archive versions that can be applied in various domains. In the following sections, we provide a detailed description of the proposed method, its implementation, and evaluation results.

2 Related Works

In recent years, the application of blockchain technology in archive version traceability and management has attract a lot of research interests. Tan et al. [4] proposed an archival data protection and sharing method based on blockchain technology, which is implemented with an interplanetary file system (IPFS) to protect, verify and share archives. Nizamuddin et al. [5] implemented a decentralized document version control with ethereum blockchain and IPFS. Albert et al. [6] proposed an novel archival system for the State archive-keeping committee of the Republic of Tatarstan (Russia). The proposed ARCHAIN is implemented with records entry consists of 8 attributes. While these studies have laid the groundwork for utilizing blockchain in archive version traceability and management, challenges such as scalability, interoperability, and regulatory compliance still need to be addressed. By building upon the insights gained from both domestic and international research endeavors, this study aims to contribute to the development of more effective and robust blockchain-based methods for managing archive versions.

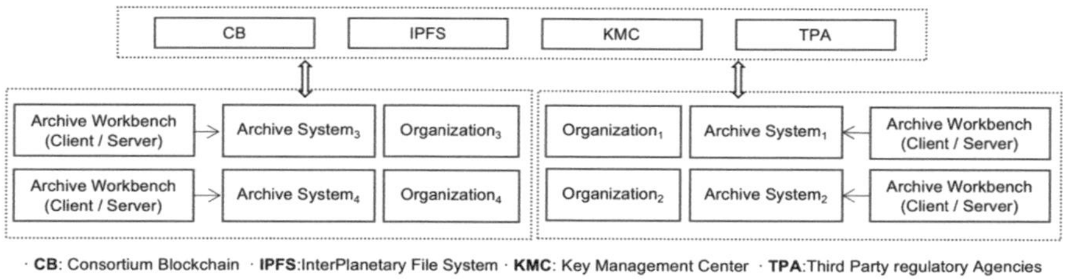

Fig. 1 The proposed architecture

3 Proposed Methods

3.1 Overview

As shown in Fig. 1, the proposed system are implemented with several key components including the Archive Information Owner (AIO), the Archival Information Visitors (AIV), Authorized Agencies (AA), Consortium Blockchain (CB), Key Management Center (KMC), InterPlanetary File System (IPFS), and Third-Party Regulatory Agencies (TPA).

In our system, we have a client/server based archive workbench, which is used for processing the collected documents. After verified and processed, the archives are transferred to the long-term preservation systems called Archive System_n in the figure. Users who are directly involved in processing and organizing electronic archival information at various stages use the proposed method to transfer archives to long-term preservation systems. In practice, they typically refer to electronic archival systems structured in a client/server model. They upload desensitized electronic archival information to the IPFS while simultaneously performing on-chain operations. There are 3 steps in the propose system including initialization, encoding/encryption and tracing. The encoding/encryption step can be implement with different environment settings. We only discuss initialization and tracing in this paper.

3.2 Initialization

The initialization step is used to generate the key for archive data. The Key Generation Center randomly selects implicit security parameters μ and a random bilinear group G, where the order of the bilinear group is a prime number q. It randomly selects an encryption exponent a, b in the order-q integer field and computes to generate the system public key PK and the master private key ZK, where $PK = \left(G, g, g^b, g^{1/b}, e(g, g)^b\right)$, $ZK = (\beta, g^a)$;

Then, the Key Management Center collects the basic attributes of all on-chain electronic archival information visitors, forming an attribute set C. Subsequently, it selects any number in the order-q integer field Z_q and c, for each attribute $C_i \in C$, randomly generates a random number $c_i \in Z_q$ to compute the private key, where $D = g^{\frac{a+c}{b}}$, $D'_{C_i} = g^{cC_i}$, $D_{C_i} = g^c \cdot H(C_i)^{cc_i}$, and H() denotes encryption using a hash function.

The key generation is performed for each AA and TPA to facilitate subsequent electronic archival encryption and storage signature operations. The public–private key pair generation process for AA and TPA is consistent. Taking the generation process for AA as an example, two large prime numbers α, β are randomly selected, and $n = \alpha\beta$, $F(n) = (\alpha - 1)(\beta - 1)$ is computed. Public key parameters e_{AA} are selected to satisfy $\gcd(e_{AA}, F(n)) = 1$. Then, inverse calculation is performed to obtain the private key exponent d_{AA} for AIV, $d_{AA} = e_{AA}^{-1}(\bmod F(n))$. This completes the generation of AA's public key A_{pub} and private keys A_{prv}, where $A_{pub} = \{e_{AA}, n\}$, $A_{prv} = \{d_{AA}, n\}$. Similarly, the public–private key pair for TPA is $T_{pub} = \{e_{TPA}, n\}$, $T_{prv} = \{d_{TPA}, n\}$.

To guarantee the distribution of organizational credentials, the TPA and AA agree on the equation $\delta = x^{token} \bmod y$, used for the identity verification of AA members AIO during the data on-chain phase. The value of δ, x, y is shared between AA and TPA. AA issues token for organizational credentials to its member AIOs.

3.3 Tracing

The abnormal operation or malicious visit of archive should be considered to trace the source and operator. Third-party regulatory agencies collaborate with authorized institutions to trace users who upload malicious, erroneous, or tampered file data. The detail implementation is described in Fig. 2.

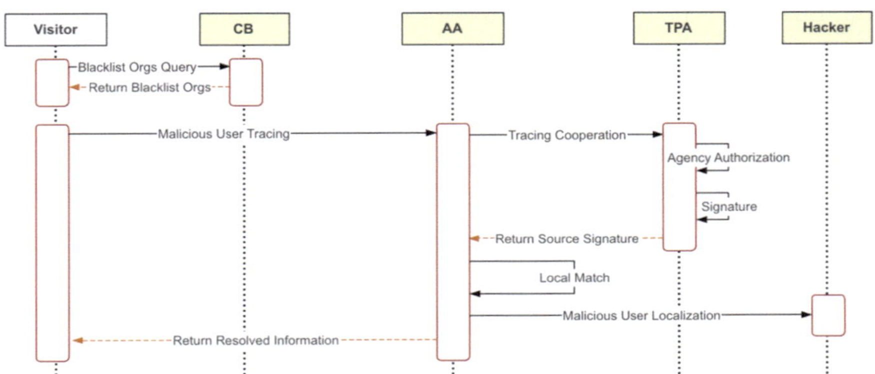

Fig. 2 The proposed tracing operation flow

(1) Trace Transaction: When users utilize the long-term storage system to retrieval archive data, the action are recorded. The long-term storage system asynchronously verifies certain blockchain data. If the verification indicates inconsistencies on the chain, the issuing institution AA of the transaction is identified via ω_{AA}, and a trace is submitted to the AO entity.
(2) Collaboration between AA and TPA: AA sends ω_{AA} and the hash value $H(M')$ of the transaction information to TPA, along with certain private key information to verify its identity. Upon receiving AA's cooperation request, TPA first validates the legitimacy of the transaction using AA's private key information. Upon successful validation, TPA calculates $\tau = (H(M')K^{e_{AA}}) \bmod n$ based on blind factor K q and then returns the result to AA.
(3) Malicious User Tracing: AA utilizes d_{AA} to calculate $\omega^*_{AA} = (\tau)^{d_{AA}} \bmod n$, computational methods to precisely locate malicious users by matching against the local database.

4 Conclusion

In this paper, we proposed a novel blockchain-based archive version traceability and management method. The proposed method can trace via chainblock platform with blind signature technology, which realized precise accountability and effective tracing in archive management in Chongqing.

References

1. Zhao, Z.: Research and design of digital archive management system based on blockchain [M.S. Thesis]. University of Science and Technology of China, Hefei (2018) (in Chinese)
2. Tschorsch, F., Scheuermann, B.: Bitcoin and beyond: a technical survey on decentralized digital currencies. IEEE Commun. Surv. Tutor. **18**(3), 2084–2123 (2016)
3. Duan, T.T., Zhang, H.W., Li, B., Song, Z.X., Li, Z.C., Zhang, J., Sun, Y.: Survey on blockchain interoperability. Ruan Jian Xue Bao/J. Softw. **35**(2): 800–827 (in Chinese) (2024). http://www.jos.org.cn/1000-9825/6950.htm
4. Tan, HB., Zhou, T., Zhao, H., Zhao, Z., Wang, W.D,, Zhang, Z.X., Sheng, N.Z, Li, X.F.: Archival data protection and sharing method based on blockchain. Ruan Jian Xue Bao/J. Softw. **30**(9), 2620−2635 (in Chinese) (2019). http://www.jos.org.cn/1000-9825/5770.htm
5. Nizamuddin, N., et al.: Decentralized document version control using ethereum blockchain and IPFS. Comput. Electr. Eng. **76**(2019), 183–197 (2019)
6. Galiev, A., et al.: Archain: a novel blockchain based archival system. In: 2018 Second World Conference on Smart Trends in Systems, Security and Sustainability (WorldS4), pp. 2018. IEEE

A Method for Predicting Smoking Population Based on Cell Phone Signaling and Questionnaire Survey

Yifeng Pi, Yingshan Cheng, Jianyong Ma, and Xu Zhang

Abstract Spatio-temporal based data analysis for population is widely studied recently. In this paper, we propose a mixed data fusion method for population prediction and smoking rating estimation. Firstly, an attention mechanism is implemented to extract human long-term stayments. Then, questionnaire data are collect via widespread survey. Finally, the data are fused for finally estimation. The method is well used in Chongqing for marking and management.

Keywords Attention mechanism · Data fusion · Population prediction

1 Introduction

Understanding and predicting population distribution in urban areas is a fundamental endeavor in demographic research and urban planning. With the development of mobile positioning technology, wireless communication technology and the popularization of smart phones provide infrastructure and hardware support for the large-scale application of mobile internet, human movement tracking, enabling users to access network services anytime and anywhere [1]. Meanwhile, with the wide application of big data technology and, people are able to more efficiently mine potential laws in massive data and accurately match similar users, thus realizing high-precision recommender services [2]. Recently, researchers have summarized the models and methods of urban population distribution estimation from different perspectives, including synthesizing the methods of population estimation. In general, spatio-temporal distribution based population estimation can be divided into two categories: static population estimation and dynamic population estimation.

Y. Pi · Y. Cheng
China Tobacco Chongqing Bureau (Company), Chongqing, China

J. Ma (✉) · X. Zhang
Department of Computer Science and Technology, Chongqing University of Posts and Telecommunications, Chongqing, China
e-mail: 2940169411@qq.com

J. S. Park et al. (eds.), *Advanced Multimedia and Ubiquitous Engineering*, Lecture Notes in Electrical Engineering 1475, https://doi.org/10.1007/978-981-95-1565-3_28

Attention network is proposed to capture spatial dependence and causal temporal convolutional network is proposed to obtain an overall temporal dependence in urban area for population prediction [4, 5]. Xu et al. [3] proposed a propose a Multi-attention based Hybrid-convolution Spatial–temporal Recurrent Network (MHSRN) for region-based traffic flow prediction, which demonstrate a good performance in region-based population estimation. In this paper, we propose a cell phone signaling and questionnaire survey based population estimation. Firstly, An attention mechanism is implemented to extract long-term stay, which can be further used for population estimation. Secondly, we design a survey paper to collect personal information, smoking behavior and related information. Finally, we process the cell phone data and questionnaire data with a data fusion method to obtain population and proportion of smoking people.

2 Method and Implementation

2.1 Long-Term Stay

It is important to classify travel preferences into long-term stable preferences and short-term dynamic preferences. Is an effective means to capture users' personalized travel characteristics, while the mining of users' long-term stable travel preferences relies on a large amount of data. As it is shown in Fig. 1, orignal user stayments are processed with attention matrix, spatio-temporal matrix and linear weight matrix to obtain a long-term stayment, which indicate the certain population in the target area.

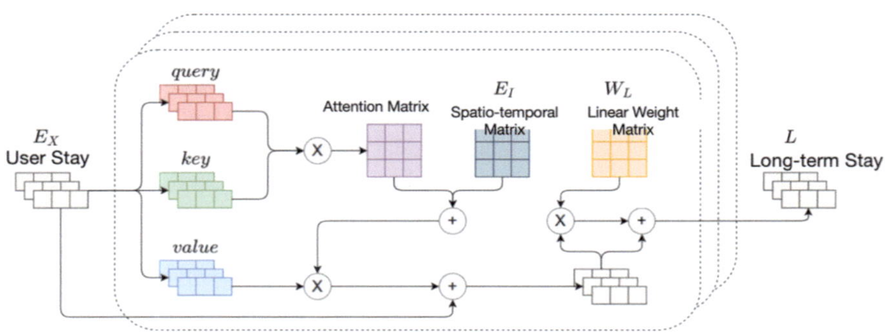

Fig. 1 The architecture of long-term stay extraction

2.2 Accumulated Human Movement and Questionnaires

Calculation of the change in human movements of a certain area from moment m to moment n in a time period of t months and weeks based on cell phone signaling:

$$\text{Move} = \left| \frac{\sum\left(\int_m^n \text{Move}_{\text{in}} - \int_m^n \text{Move}_{\text{out}}\right)}{4 \times t} \right|$$

Calculation of correction parameters for the population count at the point in time based on the trend in the degree of population movement in and out:

(1) When population movement tends to stabilize, the population estimate for that time period is relatively accurate;
(2) When the trend of population movement is large, the correction parameter for the population estimate in that time period is small;
(3) When the trend of population movement is stabilizing, the correction parameter for the population estimate within that time period increases;

The correction parameter of the population estimate at moment m is inferred from the change of the population difference in the access degree of the administrative district:

$$\beta_m = \text{Norm}\left(1 - \frac{\text{Move}_m}{\int \text{Move}}\right)$$

Calculate the number of population within the day based on the correction parameter and the handshake data pop_m at moment m:

$$\text{Population} = \sum_{m=7}^{23} \text{pop}_m \times \beta_m$$

For the questionaire survey, we select several important POIs to distribute the survey, which include personal information, trip behavior and smoking behaviors. As it is shown in Fig. 2, the long-term stay and expanded survey are processed with spatio-temporal matching and data fusion.

3 Conclusion

In this paper, we propose a mixed population prediction method, which process cell phone data and questionnaire data together. The population prediction are mainly analyzed with the attention based long-term stayments, and the smoking rating are estimated with the fusion data. The method is used well in Chongqing for marketing and management.

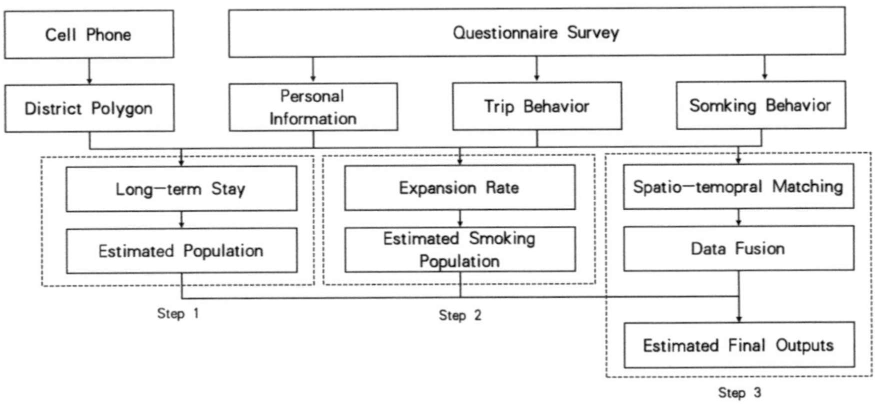

Fig. 2 Mix data processing and estimation procedure

References

1. Yu, Z.: Trajectory data mining: an overview. ACM Trans. Intell. Syst. Technol. **6**(3), 1–41 (2015)
2. Wang, J., Yang, Y., Wang, T., et al.: Big data service architecture: a survey. J. Internet Technol. **21**(2), 393–405 (2020)
3. Zhang, X., Wen, S., Yan, L., Feng, J., Xia, Y.: A hybrid-convolution spatial-temporal recurrent network for traffic flow prediction. Comput. J. **67**(1), 236–252 (2024)
4. Yang, D., Fankhauser, B., Rosso, P., et al.: Location prediction over sparse user mobility traces using rnns. In: Proceedings of the Twenty-Ninth International Joint Conference on Artificial Intelligence, Yokohama, Japan, pp. 2184–2190 (2020)
5. Liu, C.H., Wang, Y., Piao, C., et al.: Time-aware location prediction by convolutional area-ofinterest modeling and memory-augmented attentive lstm. IEEE Trans. Knowl. Data Eng. **34**(5), 2472–2484 (2020)

A Segmentation Method for Oil Debris Chains Using YOLOV7-Tiny and Labeled Watersheds

Jie Yang and Song Feng

Abstract Debris chain formation has been one of the major obstacles in the feature extraction of on-line visual ferrography. In this paper, a debris chain segmentation method based on YOLOV7-tiny is proposed. YOLOV7-tiny is used to track the growth process of the debris chain, the model output information is converted into the coordinate information of the detected wear debris in the image, which is used as the foreground markers, and the image to be segmented is thresholded binarized as the background markers, which is combined with the foreground markers, and the segmentation of the input ferrogram is performed by the mark-watershed algorithm. The method proposed in this paper is characterized by high efficiency and high accuracy, and provides a feasible segmentation method for real-time segmentation of online oil monitoring.

Keyword Debris monitoring · Debris chain segmentation · Yolov7-tiny

1 Introduction

1.1 A Subsection Sample

On-line visual ferrography (OLVF) is a new type of oil monitoring technology [1], compared with the traditional ferrography, it can help operators intuitively obtain the information of wear debris, accurately locate the type and source of wear debris, reduce the time and cost of manual intervention processing, and better grasp the operating status of the machine. OLVF for the analysis of wear debris in the fluid for the diagnosis of machinery to bring more intuitive, comprehensive, accurate, efficient and intelligent impact. By analyzing the wear debris in the oil, it can not only improve the accuracy and precision of fault prediction and diagnosis, but also better improve the reliability and operational efficiency of equipment.

J. Yang · S. Feng (✉)
Chongqing University of Posts and Telecommunications, Chongqing, China
e-mail: fengsong@cqupt.edu.cn

J. S. Park et al. (eds.), *Advanced Multimedia and Ubiquitous Engineering*, Lecture Notes in Electrical Engineering 1475, https://doi.org/10.1007/978-981-95-1565-3_29

Ferrogram segmentation plays an important role in ferrogram processing research [2]. Qiu et al. [3] proposed a K-means clustering with maximum interclass variance for effective segmentation of small wear debris, and Feng et al. [4] used contour classification filling combined with Canny algorithm to realize the segmentation of ferrograms. The principle of region-based segmentation method is to distinguish different regions by using the uniformity of the same region of the image, and finally edge pixel splicing is performed to complete the image segmentation. Wang et al. [5] used the Marked Watershed algorithm to obtain the initial segmentation of ferrograms, and then used the Gray clustering algorithm to achieve the segmentation of the deposited wear debris chain.

Deep learning based on machine learning, can carry out feature extraction and automatic learning of pattern methods, to achieve image segmentation automation and accuracy coexist, the corresponding algorithm has high robustness. For example, Feng et al. [6] constructed a Res-Unet deep network based on convolutional neural network, combined with the automatic labeling of transmitted and reflected light ferrograms to complete the segmentation of ferrograms. An et al. [7] constructed a mask R-CNN algorithm to achieve similar wear debris segmentation and recognition of complex backgrounds. Peng et al. [8] used the Inception-v3 model to automatically extract wear debris features, and the corresponding algorithm has high robustness. All these methods learn features and patterns directly from the original image without manually extracting features and setting parameters, which has high adaptive and generalization performance.

Debris chain segmentation is a prerequisite for extracting debris size distribution and morphological features. Previously, the segmentation efficiency of ferrograms was very low, and the segmentation of debris chain was based on the processing and analysis of personnel with rich experience and professional knowledge reserves. In this paper, we adopt YoLov7-tiny to track the chain formation process of wear debris deposition and combine it with the marking watershed algorithm to realize wear debris chain segmentation, thus laying the foundation for the extraction of wear debris size distribution and morphological features.

2 Proposed Method

2.1 YOLOV7-Tiny Model

The model structure of YOLOv7-tiny is shown in Fig. 1, which is a lightweight version of the YOLOv7 series, and it performs well in terms of accuracy and speed compared to the original YOLOv7 version. The model uses a generalized activation function and convolutional operation, and is able to handle the detection tasks of different types of objects, which makes it more versatile and scalable. In particular, the YOLO network model incorporates image pyramid features during the training process, which gives the YOLO network a great advantage in detecting small objects,

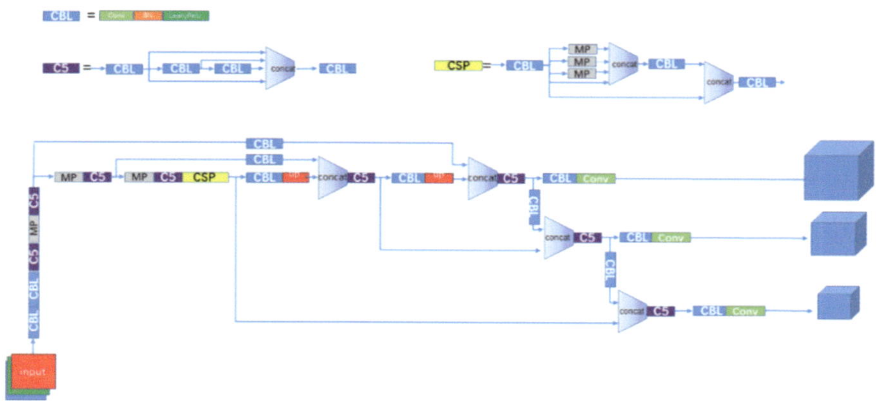

Fig. 1 Model structure of YOLOv7-tiny

which is one of the main reasons for selecting the YOLO model as the debris chain detection model. In addition, the SPPCSP network structure is used in YOLOv7-tiny, which can increase the receptive field without increasing the number of parameters, thus improving the performance of the network. Specifically, the SPPCSP network structure extracts feature maps of different sizes by using pooling kernels of different sizes, and then stitches these feature maps together to form a feature map with a larger receptive field.

2.2 Data Set

Gear wear was detected using an OLVF and videos of wear debris deposition were recorded. In the process of ferrgrams acquisition, the image quality of some of the ferrgrams is relatively poor due to the interference of air bubbles in the lubricating oil, the uneven illumination of the sensor light source, and the effect of lubricating oil contamination degradation. The quality of the training data set is important for the deep learning network model to achieve more satisfactory results. The ferrgrams with unsatisfactory imaging effect are deleted, and finally 1220 ferrgrams with good imaging effect are retained as the training data set. As shown in Fig. 2, LabelImg was used to manually label the wear debris, record the target category, location and size, and complete the production of the wear debris deposition data set.

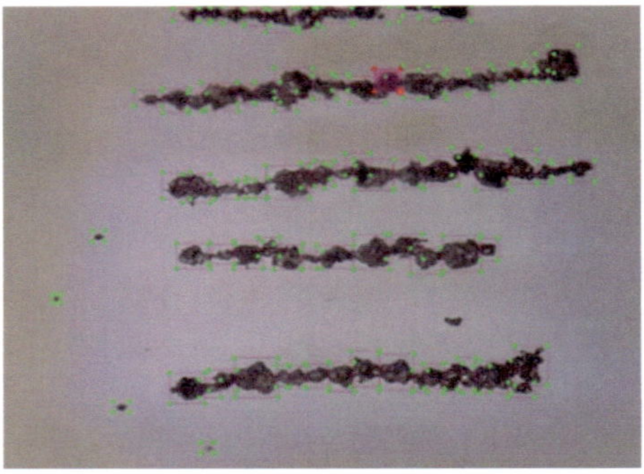

Fig. 2 Wear debris markings

2.3 Watershed Marking Algorithm

The basic idea of the watershed algorithm is to consider the map as a topographic map, where positions with higher gray values of pixels correspond to higher terrain, positions with lower gray values of pixels correspond to lower terrain, and points with higher gray values converge to form a watershed. Water is injected at the local minimum point (valley), and at the edge of the intersection of the three watersheds, the water in the three watersheds is divided into two pixels collections to complete the segmentation of the three different regions. To realize the segmentation of the image using the watershed algorithm, the pre-segmented image is first processed in response to the pre-segmented image, which is first converted into a grayscale image. In the grayscale image, these points are used as the starting point of the segmented region by finding the local minimum points and marking them as valleys. Next, the Watershed algorithm divides the pixel points into three categories: points marked as valleys, pixel points within the same segmentation region, and pixel points between different segmentation regions. Finally, the final segmentation result is obtained by merging the different segmentation regions.

3 Results and Discussions

3.1 Segmentation Results

The image to be segmented is fed into the trained YOLOV7-tiny debris chain detection model, and the position coordinates of each debris chain in the image are obtained by detection. Then, the wear debris image is subjected to gray scale extraction, binarization, morphological closure operation and expansion operation, and the background image of the image is obtained after processing, and the results are shown in Fig. 3. The image of wear debris is affected by illumination, there are light and dark areas in the background, and also a small number of noise points, after morphological processing, the influence of the environment on the segmentation is eliminated, so that the background is only black and white pixel values, in addition, It can also be found that after the image is transformed by the morphology, for the wear debris chain of the edge information features are still well preserved, and almost did not occur due to the edges will be blurred and deformation of the There is little loss of edge information due to blurring and distortion of the edges.

As shown in Fig. 4, the YOLOV7-tiny detection results are converted to coordinates in the Cartesian coordinate system, and the coordinates of the center of each debris are extracted to obtain the foreground map. Calculate the number of connected regions in the foreground map, and set a sequential growth of digital labels for each connected region as the a priori "valley" position in the mark-watershed algorithm, and then subtract the foreground position image from the background region image as the unknown position region to be determined. Finally, the segmentation is performed using the labeled watershed algorithm, and the segmentation result is shown in Fig. 5.

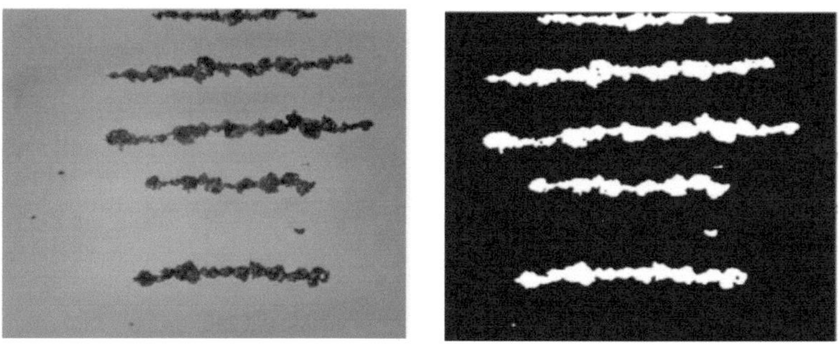

(a) Original RGB image (b) Morphological results

Fig. 3 Morphological processing of images

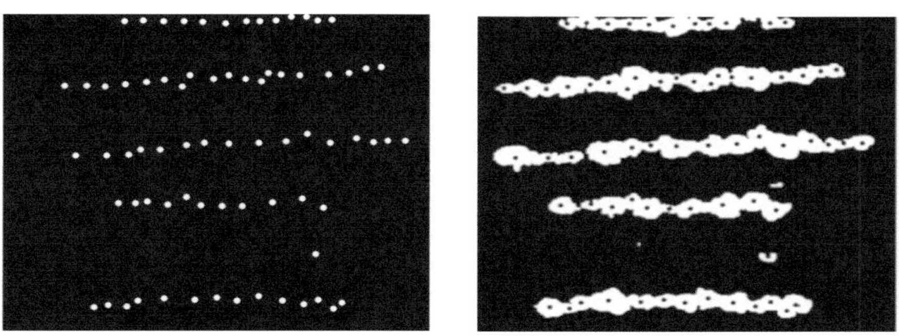

(a) Wear debris markings (b) Unknown regions to be identified

Fig. 4 Delineation of connected domains and unknown regions

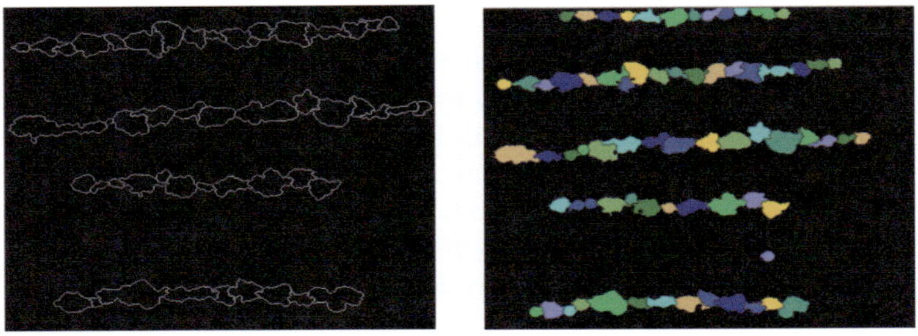

(a) Original segmentation results (b) Marked in different colors

Fig. 5 Marked in different colors

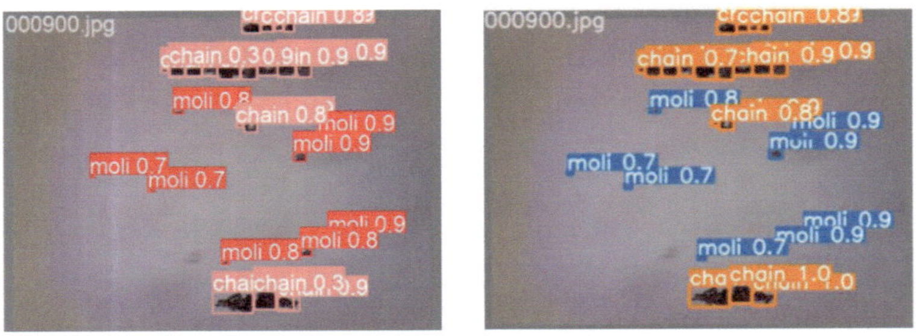

(a) Yolov5 prediction and confidence level (b) Yolov7-tiny prediction and confidence level

Fig. 6 Prediction results of the model for a batch on the test set

3.2 *Comparison with YOLOV5*

As shown in Fig. 6, it can be seen that in the place where the debris chain is piled up, the structure of the debris chain is very complex, and there are a lot of characteristics of the edge, at this time, the division of the debris chain of YOLOV5 is fuzzy, although in the setting of the lower confidence threshold, V5 is also successfully identified, but the confidence level is not high, on the contrary, the same is the complex area of debris chain piled up, V7 can be successfully identified with a higher confidence level, and then to summarize the indices listed in the above, the thesis uses the YOLOV7-Tiny model.

4 Conclusions

For the problem of debris chain segmentation in ferrogram during oil monitoring, this paper constructs a target detection network YOLOV7-Tiny, detects the debris, debris chain region in the image, obtains the coordinate information of its regional position, and realizes the debris chain segmentation by combining with the marking watershed algorithm. The method in this paper is able to segment efficiently and accurately for the fine target of debris chain. The algorithm has a high rate of correctness and robustness.

References

1. Wu, T.H., Mao, J.H., Wang, J.T., et al.: A new on-line visual ferrograph. Tribol. Trans. **52**(5), 623–631 (2009)
2. Wang, J., Bi, J., Wang, L., et al.: A non-reference evaluation method for edge detection of wear particles in ferrograph images. Mech. Syst. Signal Process. **100**, 863–876 (2018)
3. Qiu, L.J., Xuan, Z.N., Zhang, X.F.: Debris color image segmentation by K-means clustering and Ostu method. Lubrication Eng. **39**(12), 101–104 (2014)
4. Feng, S., Qiu, G., Luo, J., et al.: A wear debris segmentation method for direct reflection online visual ferrography. Sensors **19**(3), 723 (2019)
5. Wang, J., Yao, P., Liu, W., et al.: A hybrid method for the segmentation of a ferrograph image using marker-controlled watershed and grey clustering
6. Feng, S., Zeng, Q., Fan, B., et al.: Wear debris segmentation of reflection ferrograms using lightweight residual U-Net. IEEE Trans. Instrum. Meas. **70**, 1–11 (2021)
7. An, C., Wei, H., Liu, H., et al.: Ferrographic wear debris intelligent segmentation and recognition based on mask R-CNN. Lubrication Eng. **45**(3), 107–112 (2020)
8. Peng, P., Wang, J.: Wear particle classification considering particle overlapping. Wear **422**, 119–127 (2019)

ESIF-Net: Edge-Supervised Interactive Fusion Network for RGB-D Curtain Wall Frames Segmentation

Jianzhen Li, Xiaoyu Xu, Wendan Liu, Decheng Wu, Rui Li, and Sheng Liu

Abstract In this paper, we propose an edge-supervised interactive fusion network (ESIF-Net) for curtain wall frames segmentation, which utilizes deep learning techniques to adapt to the needs of curtain wall installation localization in complex building environments. Specifically, we concatenate the RGB images and the corresponding depth images in the channel dimension into the backbone network to achieve multimodal feature fusion, and introduce the Edge Feature Extraction Module (EFE) to extract edge features. Then, the Edge Features Guide Interaction Module (EGI) is designed to interact and fuse the refined features with the edge features to enhance edge contour information in the top layer. Finally, we conduct comparative experiments on the curtain wall frame dataset, and demonstrate that the proposed model has better performance. The precision, recall, accuracy, and F1-score reach 85.62%, 83.42%, 96.31%, and 84.51%, respectively.

Keywords Curtain wall frames segmentation · Multimodal feature fusion · Edge supervision

1 Introduction

At present, with the rapid development of artificial intelligence, Deep Learning (DL) has also received extensive attention and research. Among them, convolutional neural network has been widely used in automatic driving [1], face recognition [2], medical image analysis [3] and other aspects because of its powerful feature extraction ability. Recently, some scholars have applied DL to the construction field [4] to realize the automatic detection and segmentation of buildings, building materials, workers and

J. Li · X. Xu · W. Liu · D. Wu (✉) · R. Li
School of Automation, Chongqing University of Posts and Telecommunications, Chongqing, China
e-mail: wudc@cqupt.edu.cn

S. Liu
School of Data Science, Tongren University, Tongren, China

J. S. Park et al. (eds.), *Advanced Multimedia and Ubiquitous Engineering*, Lecture Notes in Electrical Engineering 1475, https://doi.org/10.1007/978-981-95-1565-3_30

so on. For example, Wu et al. [5] proposed a boundary regulated network called BR-Net for accurately segmenting roof images and extracting building contours. Wang et al. [6] used Faster R-CNN to detect nails and screws, which can assist robot patrols and achieve faster target detection.

With the rapid expansion of the technologies, automated construction has become a hot spot. For example, Han et al. [7] analyzed the process of curtain wall installation and proposed an automated system for a construction robot for curtain-wall installation (CRCWI). Through field performance testing, it can be verified that the proposed system can achieve simpler and more accurate installation, while improving the safety during installation. However, in addition to ensuring the installation speed, it is also crucial to realize area localization intelligently and efficiently. Therefore, we propose an edge-supervised interactive fusion network (ESIF-Net) for image segmentation of curtain wall frames.

2 Method

2.1 Overall Network Architecture

We propose a novel RGB-D segmentation network, called ESIF-Net, for curtain wall frame segmentation. As shown in Fig. 1. The network is an encoder-decoder structure composed of a backbone branch, EFE, EGI, and an upsampling refinement part. Two images are given, which are the RGB image of the curtain wall frame and the depth image of the corresponding scene. They are collected in the same scene, and reflect the characteristics of different aspects, with strong correlation. Therefore, we concatenate the two images in the channel dimension to form a new four-channel image, and feed it into the network. The ResNet50 network is used as the backbone network to extract multi-level RGB-D fusion features from five convolution blocks from bottom to top, denoted as $(e_1, e_2, e_3, e_4, e_5)$. Details of the key modules are described in subsequent sections.

2.2 Edge Feature Extraction Module (EFE)

The lower layers of the network can capture richer texture and boundary information. Therefore, in this section, we use the last four layers in the backbone network to induce a side branch and design an EFE module for edge feature extraction, as shown in the lower left corner of Fig. 1. Here, we discard the first layer due to its strong noise effect.

Specifically, in EFE, the given input features are first processed by $ConBR_3$ (including a 3×3 convolution layer, BN layer and ReLU layer) to reduce the channels. In addition, we use spatial attention (SA) and perform residual connections to

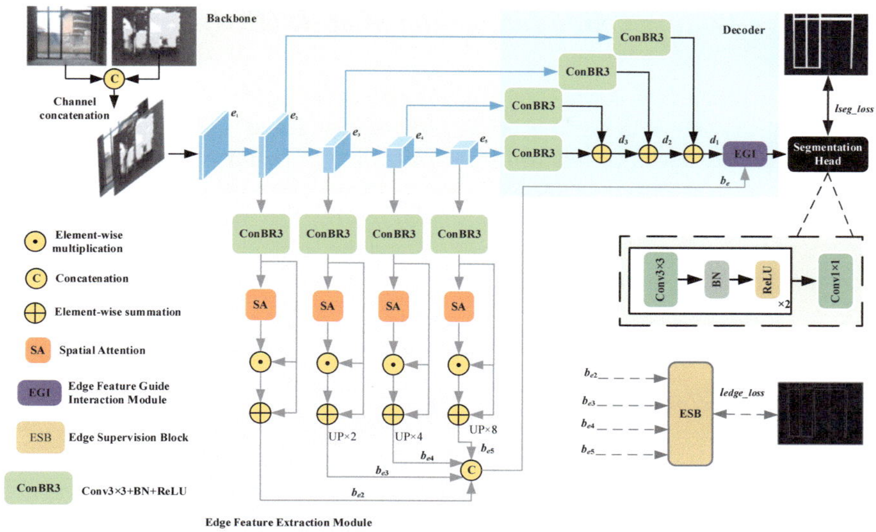

Fig. 1 Architecture of proposed ESIF-Net

realize feature enhancement, which can be defined as:

$$b_i = ConBR_3(e_i), \quad i = (2, 3, 4, 5) \tag{1}$$

$$b_{ei} = SA(b_i) \cdot b_i + b_i, \quad i = (2, 3, 4, 5) \tag{2}$$

Next, we adjust the features of all scales to the size of the second layer, and perform concatenate operation to further realize feature fusion:

$$b_e = ConBR_3(cat(b_{e2}, up_{\times 2}(b_{e3}), up_{\times 4}(b_{e4}), up_{\times 8}(b_{e5}))) \tag{3}$$

where $cat(\cdot)$ represents concatenation, $up_{\times n}(\cdot)$ denotes n times bilinear interpolation.

To capture the edge information more accurately, we also access an Edge Supervision Block (ESB) at the end of the EGE for edge supervision to capture multi-scale edge features, and use the canny operator to extract the edge from the semantic segmentation GT as the edge GT for edge detection. It can be defined as:

$$\begin{aligned} b_{edge\_out} &= Con_{1\times 1}(e_{e2}) + Con_{1\times 1}(up_{\times 2}(e_{e3})) \\ &+ Con_{1\times 1}(up_{\times 4}(e_{e4})) + Con_{1\times 1}(up_{\times 8}(e_{e5})) \end{aligned} \tag{4}$$

where $Con_{1\times 1}$ is the 1×1 convolution.

2.3 Edge Feature Guide Interaction Module (EGI)

In order to make the high-level features have richer edge features, we further introduce EGI module, as shown in Fig. 2. And we fuse the edge features obtained by the EFE and the top layer features obtained by up-sampling to perform feature interaction again. The input is first embedded into the space with channel C1 using the 1×1 convolutional layer to obtain d_q, d_k, d_v and b_q, b_k, b_v. The matrix multiplication of d_q, d_k and b_q, b_k is carried out, respectively, and the softmax function is connected to obtain the corresponding attention map, which is summarized as follows:

$$d = d_q \otimes (d_k)^T \tag{5}$$

$$d_a = soft\max(d) \tag{6}$$

$$b = b_q \otimes (b_k)^T \tag{7}$$

$$b_a = soft\max(b) \tag{8}$$

where $\otimes$ represents the matrix multiplication operation.

Next, we interactively use the attention map to weight and residually connect d_a and b_a, and then multiply with d_v, b_v respectively to capture long-distance contextual dependencies, which are summarized as follows:

$$f_b = (b \odot d_a + b) \otimes b_v \tag{9}$$

$$f_d = (d \odot b_a + d) \otimes d_v \tag{10}$$

where $\odot$ represents the pixel-by-pixel multiplication.

Finally, f_d and f_b are concatenated and added to the original input d_1 to further realize the interactive fusion of features, which is summarized as follows:

$$f_{db} = cat(f_d, f_b) + d_1 \tag{11}$$

3 Experiments and Analysis

In order to realize the intelligent segmentation of curtain wall frames, we obtain some real images of curtain wall frames which are labeled to obtain the ground truths. And the edge ground truths are obtained by Canny edge detection. The overall loss is

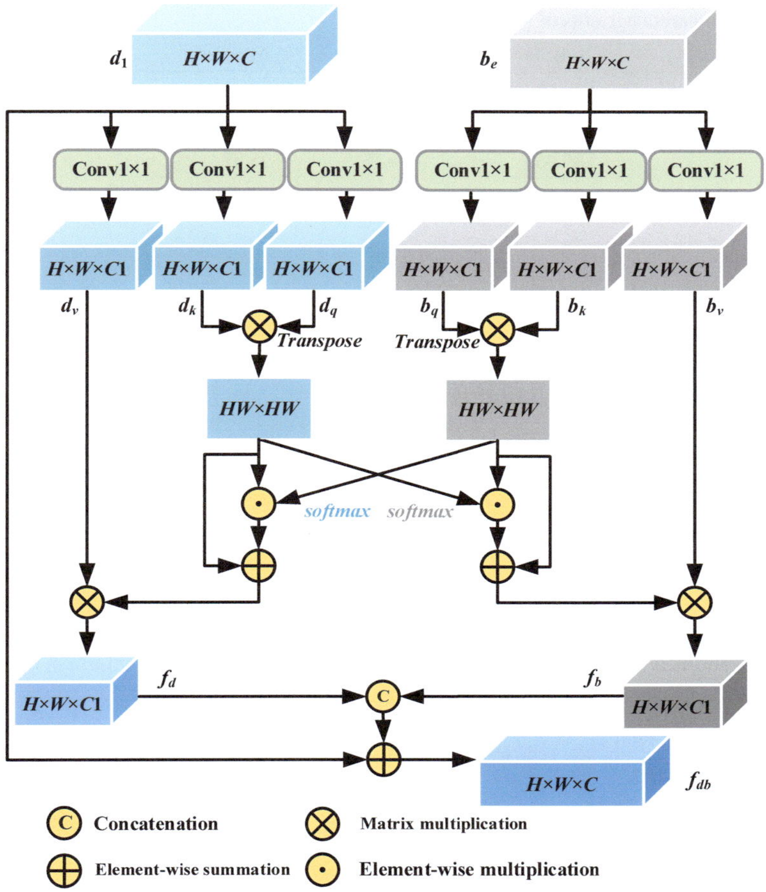

Fig. 2 Architecture of the proposed EGI

made up of segmentation loss and edge loss. Moreover, segmentation loss is used to supervise the final prediction results and edge loss is used to supervise the edge features, both of which are obtained by the binary cross entropy loss.

In addition, our network is trained by using the Pytorch toolbox on an NVIDIA GeForce RTX 4090. The initial learning rate is set to 1e-4 and 150 epochs are trained in the whole stage. In the training phase, we use data enhancement strategies for image enhancement.

To further evaluate the performance of our model, We use the precision, recall, accuracy, and F1-score as evaluation metrics to compare with some typical semantic segmentation models on the test set, including U-Net [8], Segnet [9], FCN [10], as shown in Table 1. According to the comparison of the results, we can see that our proposed model in this paper has better segmentation performance and is optimal in

Table 1 Comparative results of metrics on the test set

Models	Precision	Recall	Accuracy	F1-score
U-Net	79.32	72.69	94.35	75.86
Segnet	73.99	67.54	93.13	70.62
FCN	85.45	79.77	95.87	82.51
ESIF-Net (Ours)	**85.62**	**83.42**	**96.31**	**84.51**

Table 2 Results of ablation experiments

Models	Precision	Recall	Accuracy	F1-score
Baseline	83.94	81.35	95.87	82.62
Baseline + EFE	85.11	82.92	96.19	84.00
Baseline + EFE + EGI (Ours)	**85.62**	**83.42**	**96.31**	**84.51**

all evaluation metrics, and the precision, recall, accuracy, and F1-score reach 85.62%, 83.42%, 96.31%, and 84.51%, respectively.

To verify the contribution of the key modules to our model, we conduct ablation experiments, as shown in Table 2. From the results, it can be seen that the proposed model with the introduction of EFE and EGI achieves better results than Baseline, and each module contributes differently to the segmentation performance.

4 Conclusion

In this paper, we propose an edge-supervised interactive fusion network (ESIF-Net) for RGB-D curtain wall frames segmentation. In particular, EFE and EGI are introduced to capture multi-scale edge contour information and further enrich the edge features in high-level features. In order to demonstrate the effectiveness of our method, we compare our method with the typical segmentation models, and the experimental results show that our method achieves better performance on the four evaluation metrics of the precision, recall, accuracy, and F1-score.

Acknowledgements This work was supported by the National Key Research and Development Program of China (2022YFE0204500), the Chongqing Special Key Project for Technological Innovation and Application Development (cstc2021jscx-gksbX0055, CSTB2022TIAD-KPX0134), the Science and Technology Research Program of the Chongqing Municipal Education Commission (KJQN202200637), Chongqing Talent Plan(CQYC202203091156), and the Innovation Research Group of Universities in Chongqing (CXQT20016).

References

1. Huang, Z., Lv, C., Xing, Y., Wu, J.: Multi-modal sensor fusion-based deep neural network for end-to-end autonomous driving with scene understanding. IEEE Sens. J. **21**(10), 11781–21179 (2021)
2. Dwijayanti, S., Iqbal, M., Suprapto, B.Y.: Real-time implementation of face recognition and emotion recognition in a humanoid robot using a convolutional neural network. IEEE Access **10**, 89876–89886 (2022)
3. Niyas, S., Pawan, S.J., Kumar, M.A., Rajan, J.: Medical image segmentation with 3D convolutional neural networks: a survey. Neurocomputing **493**, 397–413 (2022)
4. Jacobsen, E.L., Teizer, J.: Deep learning in construction: review of applications and potential avenues. J. Comput. Civ. Eng. **36**(2), 03121001 (2022)
5. Wu, G., Guo, Z., Shi, X., Chen, Q., Xu, Y., Shibasaki, R., Shao, X.: A boundary regulated network for accurate roof segmentation and outline extraction. Remote Sens. **10**(8), 1195 (2018)
6. Wang, Z., Li, H., Zhang, X.: Construction waste recycling robot for nails and screws: computer vision technology and neural network approach. Autom. Constr. **97**, 220–228 (2019)
7. Han, C.S., Lee, S.Y., Lee, K.Y., Park, B.S.: A multidegree-of-freedom manipulator for curtain-wall installation. J. Field Robot. **23**(5), 347–360 (2006)
8. Ronneberger, O., Fischer, P., Brox, T.: U-net: convolutional networks for biomedical image segmentation. In: International Conference on Medical Image Computing and Computer-Assisted Intervention. Springer, pp. 234–24 (2015)
9. Badrinarayanan, V., Kendall, A., Cipolla, R.: Segnet: a deep convolutional encoder-decoder architecture for image segmentation. IEEE Trans. Pattern Anal. Mach. Intell. **39**(12), 2481–2495 (2017)
10. Long, J., Shelhamer, E., Darrell, T.: Fully convolutional networks for semantic segmentation. In: IEEE Conference on Computer Vision and Pattern Recognition, pp. 3431–3440 (2015)

Research on Mechanical Properties of the Bolted Joint of an EHV Tower for Smart Transmission Line

Mengyang Shui, Yang Zhao, Baojun Yang, and Sheng Lu

Abstract Transmission towers are the main link of electric energy transmission, exchange and regulation in the power system. The bolted joints play a vital role in ensuring the bearing capacity of the transmission towers. To realize the smart transmission line and build the database of the hysteresis performance of the tower, it is necessary to dig into the mechanical properties of the bolted joint. In this paper, two types of finite element models of a 500 kV wine-glass extra high voltage (EHV) transmission tower were set up by ABAQUS to assess the dynamic response under wind load. Then the hysteresis curve and energy dissipation coefficient of a typical bolt joint connection with different displacement amplitudes were explored by numerical simulation. After that, the influence of the friction coefficient on the mechanical properties was analyzed. The results show that the displacement amplitude has a great influence on the hysteresis performance. The energy dissipation coefficients of the joints fluctuate with displacement amplitudes and reach a peak at 1.4 mm. Increasing the friction coefficient of the contact surfaces can enhance the deformation capacity. The research can provide a reference for the design and optimization of the bolted joint connection of the tower and lay a foundation for digital twins of future transmission lines.

Keywords Smart transmission line · Bolted joint · Mechanical property · Hysteresis performance

M. Shui · Y. Zhao (✉) · S. Lu
School of Advanced Manufacturing Engineering, Chongqing University of Posts and Telecommunications, Chongqing, China
e-mail: zhaoyang@cqupt.edu.cn

B. Yang
Chongqing Huashu Robotics Co., Ltd., Chongqing, China

J. S. Park et al. (eds.), *Advanced Multimedia and Ubiquitous Engineering*, Lecture Notes in Electrical Engineering 1475, https://doi.org/10.1007/978-981-95-1565-3_31

1 Introduction

Electric energy is the main energy and power widely used in various fields such as modern industrial and agricultural production, scientific and technological research. Driven by the triple goal of carbon peaking and carbon neutrality targets, the national digital economy development strategy, and domestic energy demand, China is accelerating energy transformation and promoting the construction of a new energy-based power system. By taking big data as an essential support, the cloud computing, blockchain, artificial intelligence, and other information technology are used to establish a digital twin model of the power system. And the model can be used for intelligent data analysis and making decision, real-time monitoring of the state of grid operation, and predicting the future development trend of the grid. It is crucial to ensure that the grid is economical and stable operation. As the core parts of overhead transmission lines, the carrying capacity of the transmission towers plays a crucial role in the normal operation of the power system [1]. The transmission tower is a kind of tall flexible structure, which is made by assembling multiple sheets of angle steel and bolts. The bolts may come loose under the cyclic load, such as strong wind and de-icing vibration. The looseness of bolts may lead to the falling off of the nuts and the reduction of the bearing capacity, even bringing down the towers [2]. Therefore, it is of great significance to grasp the mechanical properties of the bolted joint to prevent the bolts from loose and ensure the safety of the transmission tower.

At present, scholars have conducted a large number of studies on the statics mechanical properties of bolted connection joints of transmission towers. Tan studied the joint parameters of bolted joints' influence on the hysteresis behavior by the bolt analytical and finite element model [3]. Park also verified two types of friction-type reinforcing members to upgrade the wind resistant performance of transmission towers by cyclic loading tests [4].

Transmission towers are in a strongly nonlinear state under wind loads, and the dynamic response analysis relying on the traditional finite element method still cost a lot of time. The neural network method with some valid samples can directly establish the relationship between the load excitation and the structural response to predict the wind response of the structure. Therefore, in this paper, two types of finite element models of a 500 kV wine-glass EHV transmission tower were set up by ABAQUS to assess the dynamic response under wind load. Then the hysteresis curve and energy dissipation coefficient of a typical bolt joint connection were explored by numerical simulation. After that, the influence of friction coefficient on mechanical properties was analyzed. On this basis, developing a database of hysteresis curves and training in neural networks, then the real-time feedback of the hysteresis performance of bolted connection can be realized, which lays a fundamental foundation for digital twins of transmission lines.

The remainder of the paper is organized as follows: Sect. 2 establishes the model of the transmission tower and the typical bolted joint. Section 3 discusses the parameters that influence the properties of the bolted joint. Section 4 summarizes the paper.

2 The Establishment of the Finite Element Model

2.1 Comparison Between Beam-Truss Hybrid Model and Partial Solid Model

A 500 kV wine-glass transmission tower, which is located in the middle of two-span 400–400 m conductor wires is selected as the research object. A three-dimensional (3D) beam and truss hybrid model of the tower and the cable element to model the lines are set up in reference [5] (see Fig. 1a). After computation of the dynamic response of the steel tower under the fluctuating wind, it can be found that the dangerous zone appears in the main material of the tower body. A partial solid model that considers the real bolted joint is adopted to study the bearing capacity under wind load. The established solid model of the components in the dangerous area consists of five sheets of angle steel and thirty M20 bolts (see Fig. 1b). The finite element simulation analysis of these two models under the equivalent static load is carried out respectively. And the design reference wind speed is 30 m and the load applied on the lines and tower is converted into the form of acceleration. Figure 1a shows the results of the 3D beam and truss hybrid model. It can be seen that the maximum Mises stress is 180.6 MPa. Figure 1b shows the results of the partial solid model of the tower after removing the bolts and angle steels, and its maximum stress is 188.4 MPa. By comparing the two finite element results, it can be found that the calculation results of the tower's overall stress in the partial solid model are very close to those of the beam and truss hybrid model. Furthermore, the maximum stress solid part appears near the bolt hole of the angle steel, as shown in Fig. 1b. The reason is the excessive stress caused by the contact between the bolt and the bolt hole under wind load.

According to the above analysis, it can be seen that the bolted joint which connects the bolt and angle steel is a vital component in the carrying capacity of the tower. Therefore, it is necessary to study further the mechanical property of the bolt connection section.

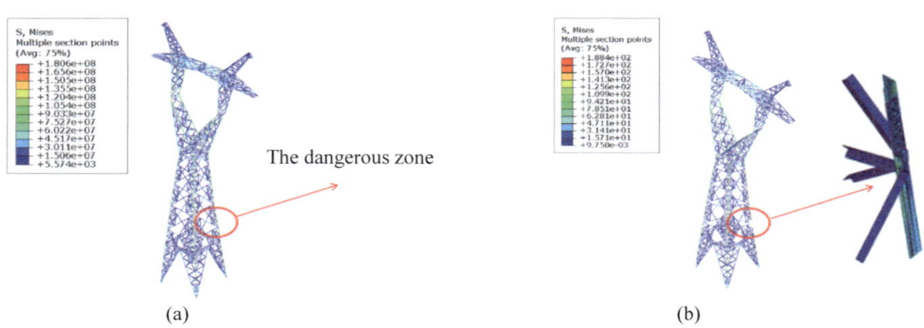

Fig. 1 Two types of finite element models of a tower. **a** 3D beam and truss hybrid model; **b** partial solid model

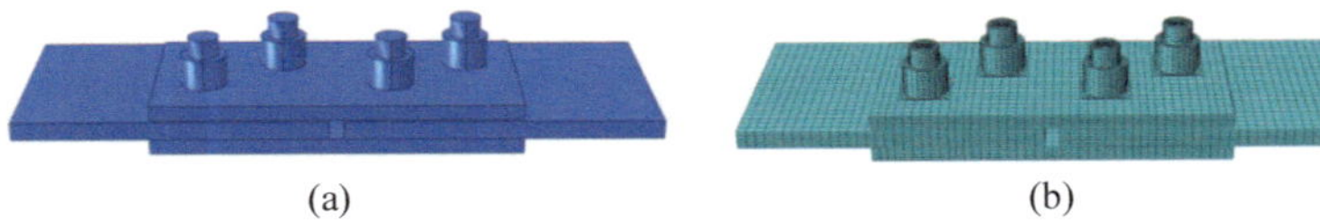

Fig. 2 Finite model of bolted connection structure. **a** Geometric model; **b** Meshing

Table 1 Material parameters of the bolted joint

Property	Plate	Bolt
Material	Q420	45#
Young's modulus	206GPa	212GPa
Poisson's ratio	0.28	0.291
Density	7.85×10^3 kg/m^3	7.87×10^3 kg/m^3

2.2 Finite Element Model of Bolted Joints

To study the hysteresis performance of bolt joints, a 3D geometric model of the bolt connection structure of the 500 kV wine glass transmission tower is created in ABAQUS. The geometric model and the corresponding mesh are shown in Fig. 2. The material properties of the plate and the bolt are shown in Table 1.

The pre-tightening force of the bolts is applied in ABAQUS as bolt load. According to the design codes of the transmission tower and the type of bolts, the force is set as 27 kN. The boundary condition of the finite element model is restricting all of the degrees of freedom of the left end surface of the middle connecting plate. And the right end surface of the middle connecting plate is subjected to a low-cycle displacement in the X direction (see Fig. 3). Then the load–displacement curves of the bolted joint under cyclic load can be obtained through numerical computation.

3 The Friction Coefficients Influence on Properties of the Bolted Joint

3.1 The Hysteresis Curves

The hysteresis curve is the function between the cyclic force and the deformation under the action of the force. If the annular area enclosed by the hysteresis curve is large, it indicates that the whole structure has good deformation capacity and the ability to absorb vibration energy. After many computations, the hysteresis curves under various friction coefficients are plotted in Fig. 4. The hysteresis curve is a straight line starting from the origin of the coordinates. The reason is that the bolted joint at the beginning stage is in an elastic state due to the small displacement load. With the increase of displacement, the bolted joint begins to slip, and the hysteresis

Fig. 3 The displacement amplitude curve

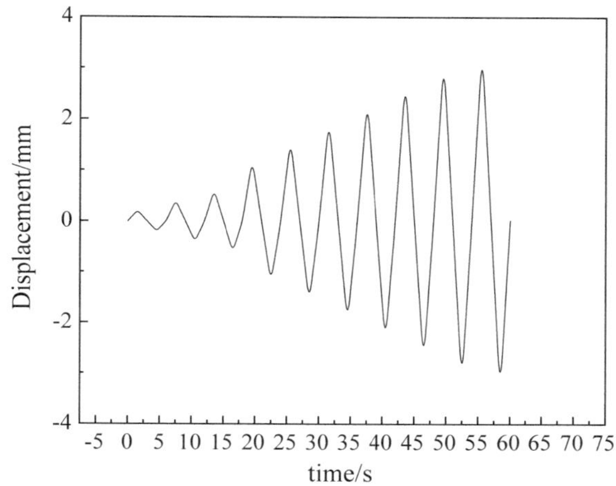

loop area formed by the curve becomes larger and larger. It means that the bolted joint can resist vibration caused by external load well and absorb vibration energy to reduce vibration intensity effectively. As the displacement amplitude increases continually, the bolted joint produces a second slip which appears between the upper and lower connecting plates. The hysteresis curve gradually takes on a z-shaped and pinched shape as the amplitude of the end position shift increases to 3.15 mm. It can be seen from the results that the shapes of the hysteresis curves are almost the same, but the area of the hysteresis loop formed by the hysteresis curve is larger with the friction coefficient increasing. The results show that the deformation capacity of bolted joints is enhanced by increasing the friction coefficient of each contact surface.

3.2 *The Energy Dissipation Coefficient*

The energy dissipation coefficient is often introduced to study the energy dissipation capacity of the joint, and the energy dissipation coefficient can be expressed as (1).

$$E = \frac{S_1}{S_2} \tag{1}$$

where S_1 indicates the energy loss by the whole structure under cyclic displacement loading. S_2 represents the energy absorbed by the structure during the whole elastic deformation stage.

The energy dissipation coefficients of joints under different friction coefficients are illustrated in Fig. 5. In the process of displacement load increasing, the energy dissipation coefficient of the joint shows a general trend of increasing first, then

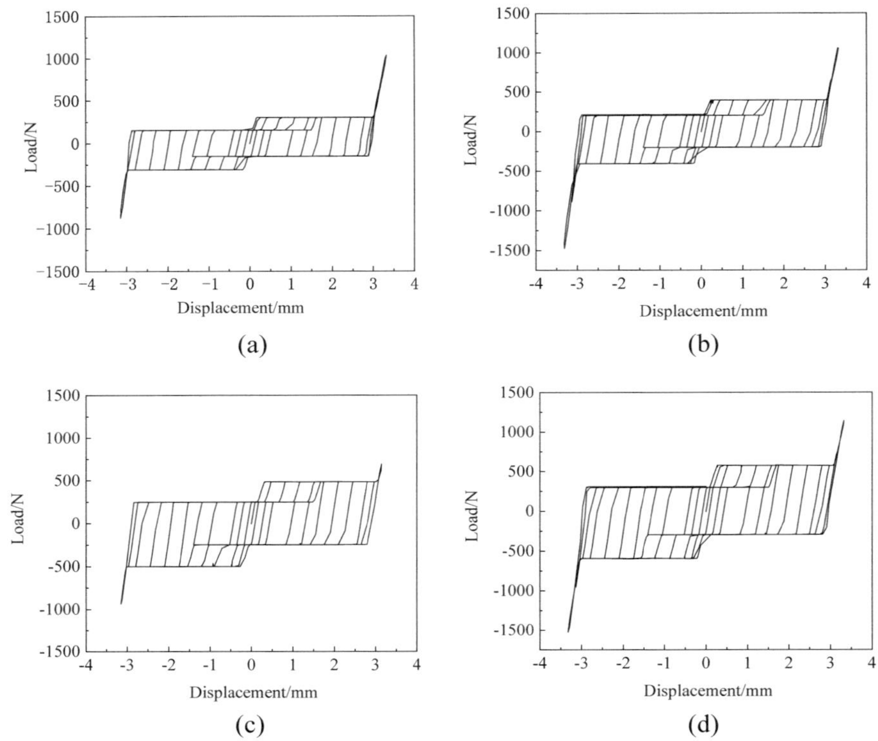

Fig. 4 The hysteresis curves under different friction coefficients. **a** $f = 0.16$; **b** $f = 0.21$; **c** $f = 0.26$; **d** $f = 0.31$

decreasing to a certain extent, then increasing again, and finally decreasing. When the displacement amplitude is equal to 1.4 mm, the energy dissipation coefficient reaches the maximum value. At this point, the joint has the strongest energy dissipation capacity. As the friction coefficient increases, the joint energy consumption coefficient keeps the same trend as before, which indicates that the friction coefficient has little effect on the energy dissipation coefficient of the joint.

4 Conclusions

The bolted joint structure of a 500 kV EHV transmission tower is chosen as the research object in this research. The finite element models of the tower and bolted joint are established by ABAQUS, and the mechanical properties of the bolted joint are obtained. The main conclusions are as follows:

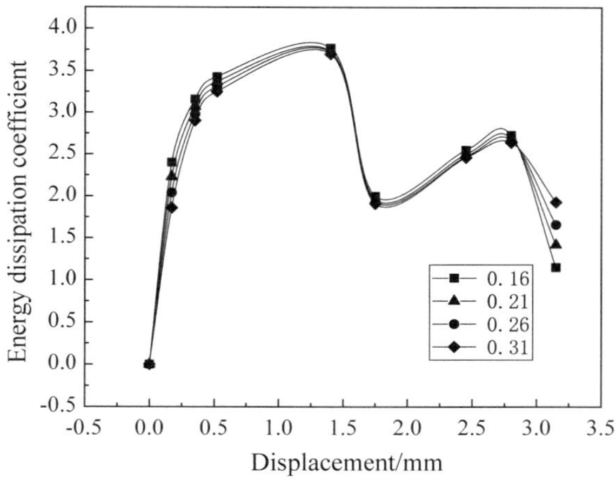

Fig. 5 The influence of friction coefficients.

Conclusions

(1) The stress response under wind load of the partial solid model and the beam and truss hybrid model has a little different. Excessive stress may appear in the solid of the bolt and bolt hole.
(2) The displacement amplitude has a great influence on the hysteresis performance of bolts. With the increase of displacement amplitude, the hysteresis curve gradually shows a Z-shaped shape. The energy dissipation coefficients of the joints fluctuate with displacement amplitudes and reach a peak at 1.4 mm.
(3) The deformation capacity of the bolted joint is enhanced by increasing the friction coefficient of each contact surface. And the friction coefficient has little effect on the energy dissipation coefficient of the joint.

The methods and results introduced in this paper can gain greater insight into the hysteresis performance of the bolted joint under cyclic load and offer a theoretical basis for the dynamic analysis of the transmission tower.

References

1. Vinay Kumar, S.: Analysis of transmission line tower subjected to wind loading. Mater. Today: Proc. 1739–1747 (2022)
2. Swabarna, R.: State of the art review of wind induced vibration and its control on transmission towers. Structures **29**, 254–64 (2021)
3. Tan, L.: Study on hysteresis and threaded fitting behavior of bolted joint with non-parallel bearing surface. Mech. Syst. Signal Process. **168**, 108655 (2022)
4. Ji-Hun, P.: Cyclic loading test of friction-type reinforcing members upgrading wind-resistant performance of transmission towers. Eng. Struct. **29**, 3185–96 (2007)
5. Wu, C.: Prediction on coupled transmission tower-line system dynamic response under strong wind based on support vector regression. Energy Rep. **8**, 260–269 (2022)

Medical Robot for Traditional Chinese Medicine Based on Infrared Thermal Imaging

Xingchen Liu, Xiaoyu Xu, Rui Li, Decheng Wu, Shiming Wu, Jiahua Qin, and Na Luo

Abstract Traditional Chinese medicine (TCM) treatments such as moxibustion and massage are often plagued by challenges such as imprecise positioning, and labor intensiveness. Hence, we propose an innovative solution by integrating infrared thermography and robotics into TCM practices. With our approach, doctors can select the treatment points on a patient's thermal imaging, which are then accurately mapped onto the visible image of the patient through key point detection and coordinate mapping. Subsequently, the robot is precisely controlled to administer treatments such as moxibustion and massage on these points. The experimental results show that this TCM medical robot can operate stably during the treatment process and can complete the tasks of positioning and treatment.

Xingchen Liu, Xiaoyu Xu—These authors contributed equally to this work.

X. Liu · X. Xu · R. Li · D. Wu (✉)
Chongqing University of Posts and Telecommunications, Chongqing, China
e-mail: wudc@cqupt.edu.cn

X. Liu
e-mail: lxchennn@163.com

X. Xu
e-mail: 13210086732@163.com

R. Li
e-mail: lirui@cqupt.edu.cn

S. Wu · J. Qin · N. Luo
Chonggang General Hospital, Chongqing, China

Chongqing Baotonghua Medical Technology Company Limited, Chongqing, China

S. Wu
e-mail: 1758wsm@163.com

J. Qin
e-mail: 3473606992@qq.com

N. Luo
e-mail: 601064638@qq.com

J. S. Park et al. (eds.), *Advanced Multimedia and Ubiquitous Engineering*, Lecture Notes in Electrical Engineering 1475, https://doi.org/10.1007/978-981-95-1565-3_32

Keywords Infrared thermal imaging · TCM medical robotics · Moxibustion · Key point detection · Coordinate mapping

1 Introduction

In recent years, infrared thermography has been widely used in painful diseases and early cancer screening. Combining infrared thermography with the idea of treating the unhealed in Chinese medicine can provide more early and accurate treatment services. Currently, TCM treatment has problems such as high specialization and time-consuming, which leads to lower treatment efficiency.

With the development of science and technology, combining TCM treatment with robotics effectively improves the efficiency of TCM treatment. Reference [1] introduced an advanced robotic surgical platform that provides minimally invasive techniques in a variety of complex clinical surgical procedures. Reference [2] developed a TCM massage robot through robot vision technology for the identification and localization of acupuncture points. However, there is a lack of TCM medical robots that combine with infrared thermography to accurately and efficiently combine infrared thermography with TCM treatment modalities. To solve this problem, we introduce a medical robot system for TCM based on infrared thermography. This system locates the points on the infrared thermography of the human body directly to the real human body and can move the end of the machine to the target points to carry out therapeutic operations such as moxibustion and massage.

2 Overall Scheme Design

The research on the TCM medical robot system based on infrared thermal imaging is grounded in the principles of machine vision. The integral components of this robotic system comprise a three-axis coordinate robot, a visible light camera, three controllers, three actuators, a brake, a range sensor, a temperature sensor, and a pressure sensor, among others. A schematic representation detailing the overall program flow is illustrated in Fig. 1.

Step (1), the infrared thermal image of the patient is acquired and key point detection is performed on the infrared thermal image. Step (2), the visible image of the patient is acquired and key point detection is performed on the visible image. Step (3), the doctor selects the target points (one or more) to be treated on the infrared thermal image, and the system maps the target points onto the visible image. Step (4), the target point position information is sent to the robot control system to control the robot movement.

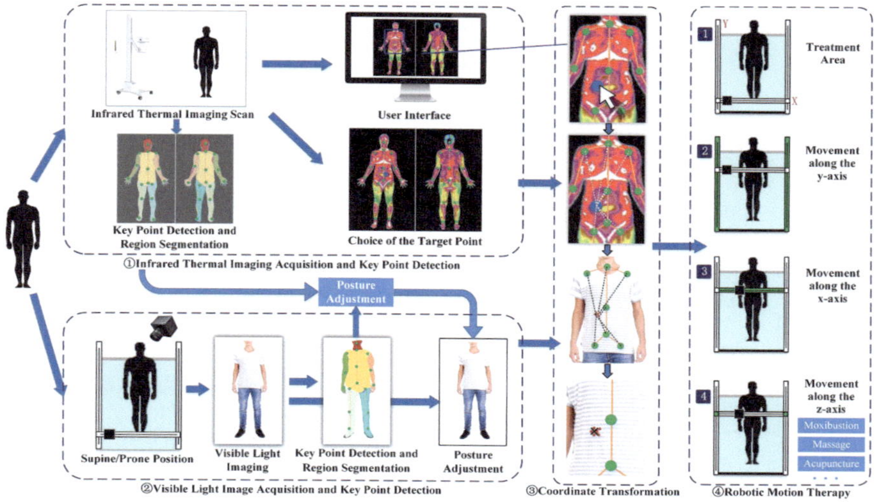

Fig. 1 The comprehensive program flow of TCM medical robots

3 Robot Software Design

In this section, we describe the software part of the TCM medical robot. The target point matching and localization of the TCM medical robot needs to be done using algorithms, i.e., matching the infrared thermal image of the patient with the visible light image, and localizing the target point from the infrared thermal image to the visible light image. The software part involves algorithms such as human key point detection and coordinate mapping. The flowchart of the algorithm is shown in Fig. 2.

We used the format of the COCO2017 dataset as an annotation and labeled 17 key points for the human body, such as the head, shoulders, wrists, elbows, hips, knees, and ankles, among other key areas. Keypoint RCNN combined with Residual Network (ResNet) [3] and Feature Pyramid Network (FPN) [4] is used to detect human infrared thermography and visible light images, the output of the network results in multiple individuals and multiple alternative detection frames and after filtering, we get the unique optimal output.

With the key point detection results, we regionally divide the human body in infrared thermography into five parts: left arm, right arm, torso, left leg and right leg. Then, we find out the region where the target point to be treated is located, and judge the location of the selected point in the order of torso, left arm, right arm, left leg and right leg. If the target point is located in this region of the human torso, the geometric relationship is established between the target point and the key points in this region, the geometric relationship is mapped onto the visible image according to the scaling ratio, and the target point to be treated is found on the visible image by the fitting algorithm.

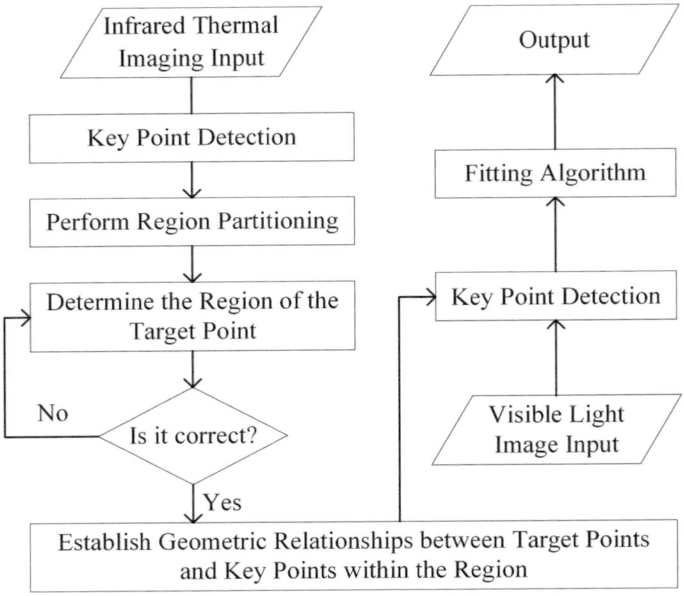

Fig. 2 The algorithmic flowchart

4 Robot Hardware Design

In the overarching scheme delineated in Sect. 2, we provided a concise overview of the robot's overall architecture. In this subsequent section, we embark on a detailed exposition of the diverse hardware components along with their respective installation sites. The hardware configuration is visually depicted in Fig. 3.

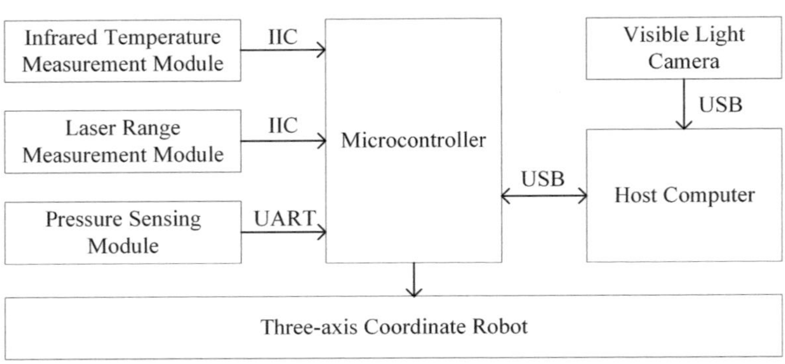

Fig. 3 Diagram of hardware architecture

We have chosen a three-axis coordinate robot, which is driven by three stepper motors. One of them is equipped with a brake in the z-axis direction to ensure that the machinery remains smooth and reliable during operation. The controller is controlled by a microcontroller and the sensors are communicated with the microcontroller through serial ports or IIC. We chose the infrared laser distance measuring module, this laser distance measuring sensor is installed to the end device of the machine, in the process of descending the end device of the machine, the laser distance measuring sensor always scans the distance from the patient's body, according to the different treatment modalities, set up different descending distance. We also choose the infrared temperature measurement module, the infrared temperature measurement sensor will be installed to the machine end device, if the treatment method uses moxa sticks and other objects with heat source, then the patient's skin area of the treatment to measure the temperature, to avoid the temperature is too high for the patient to cause harm.

5 Results of the Experiment

In this section, we will explore the utilization of moxibustion treatment as a prime example. Our approach involves securing moxa strips onto the robot's end effector using a specialized moxa clamping device. The deployment of the robot, including its moxa device, control system, and other components, enables the seamless execution of moxibustion treatment operations.

The experimental results showed that the robot was able to move to the target point position according to the instruction with a localization error of about 1.5 cm, and the robot's end device was able to maintain a safe distance from the human body. After a certain time or temperature has been reached for treatment at one point, the robot is able to run to the next point for treatment. The experimental results are shown in Fig. 4. The robot can realize accurate point finding and move according to the planned path in the real environment, and the running process is smooth and safe.

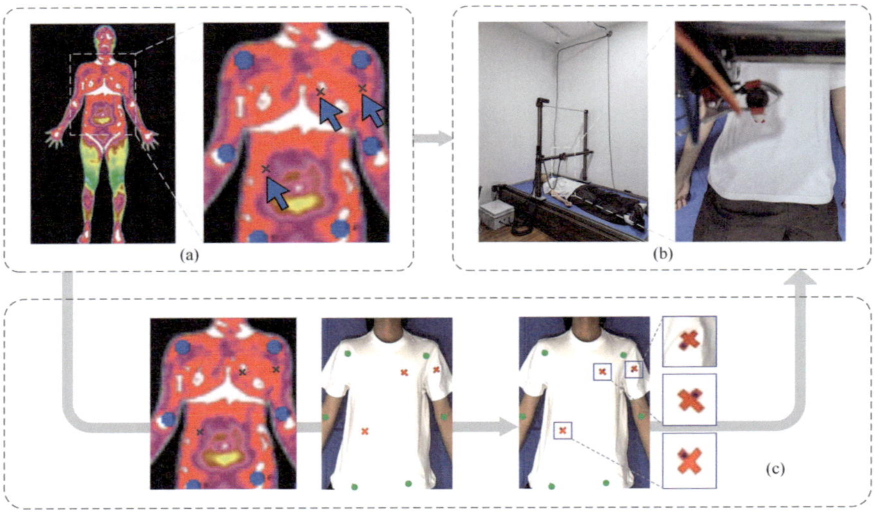

Fig. 4 Results of the experiment. **a** represents the infrared thermography part. **b** represents the physical part of the machine. **c** represents the experimental part, where the red crosses represent the marked points and the blue points above the red crosses represent the calculated results

6 Conclusion

In this paper, we propose a TCM medical robot system based on infrared thermography. The system can map the points on the patient's infrared thermal imaging to the patient's real human body by combining hardware and software, and perform robotic moxibustion, massage and other TCM treatments. Experiments show that this TCM medical robot can operate stably during the treatment process and can successfully complete the tasks of localization and treatment. In the future work, we plan to focus on improving the algorithm accuracy as well as robot hardware optimization.

Acknowledgements This work was supported by the Chongqing Special Key Project for Technological Innovation and Application Development (CSTB2022TIAD-KPX0134), the Science and Technology Research Program of the Chongqing Municipal Education Commission (KJQN202200637), Chongqing Talent Plan(CQYC202203091156), the Innovation Research Group of Universities in Chongqing (CXQT20016), and the Cooperation Project between Universities in Chongqing and Institutes Affiliated with the Chinese Academy of Sciences (HZ2021018).

References

1. Li, J., Fang, A.: Overview and clinical applications of the da vinci intelligent surgical robot. China Med. Device Inf. **25**(16), 32–33 (2019)

2. Gao, H., Lu, S., Wang, T.: Research and development of Chinese medical massage robot. ROBOT **33**(5), 553–562 (2011)
3. He, K., Zhang, X., Ren, S.: Deep residual learning for image recognition. In: Proceedings of the IEEE Conference on Computer Vision and Pattern Recognition. IEEE, Las Vegas, pp. 770–778 (2016)
4. Lin, T.Y., Dollár, P., Girshick, R.: Feature pyramid networks for object detection. In: Proceedings of the IEEE Conference on Computer Vision and Pattern Recognition. IEEE, Honolulu, pp. 2117–2125 (2017)

A Sparse Array with Increasing Spacing for Fourth-Order Different Co-array

Jing Zhao, Sheng Liu, and Decheng Wu

Abstract In this paper, a sparse array with increasing spacing is proposed to construct a fourth-order different co-array (FODCA) for direction of arrival (DOA) estimation. Being different from the conventional sparse arrays, the spacing of adjacent sensors gradually increases according to an arithmetic sequence. The element spacing between the two closest sensors is many times the half-wavelength of the incident signal. Hence, the proposed array can reduce the mutual coupling effect more effectively than many other types of sparse arrays. Some simulation results can prove the effectiveness of the proposed array.

Keywords Increasing spacing · Fourth-order co-array · Mutual coupling

1 Introduction

Recently, sparse arrays have been widely used in the direction of arrival (DOA) estimation of spatial signals. Compared with the uniform arrays, sparse arrays have larger element spacing and various array structures. Hence, the sparse arrays are less affected by the mutual coupling effect and can provide more degrees of freedom (DOFs) by using the co-array concept.

Nested array [1] and co-prime array [2] are two kinds of sparse arrays widely studied in the field of array optimization. Some modified co-prime arrays [3, 4] and nested arrays [5–8] also have been proposed on the basis of the sparse arrays [1, 2]. For these sparse arrays, using the continuous virtual sensors in the second-order difference co-array (SODCA), the extended covariance matrices can be constructed. By dealing with the extended covariance matrices, more signals can be resolved and more accurate DOA estimates can be obtained.

J. Zhao · S. Liu (✉)
Tongren University, Guizhou, China
e-mail: dshls@gztrc.edu.cn

D. Wu
Chongqing University of Posts and Telecommunications, Chongqing, China

J. S. Park et al. (eds.), *Advanced Multimedia and Ubiquitous Engineering*, Lecture Notes in Electrical Engineering 1475, https://doi.org/10.1007/978-981-95-1565-3_33

In fact, the fourth-order cumulant-based DOA estimation algorithm can resolve more signals than the second-order statistics-based DOA estimation algorithm. By extending the conventional nested array and co-prime array, the four-level nested array [9] and the extended co-prime array [10] were proposed based on the fourth-order cumulant. Subsequently, an extended nested array for four-order cumulant was proposed in [11] and it can obtain more continuous fourth-order lags than [9, 10]. In [12], an expanding and shift-based (ESA-based) scheme was proposed to extend the nested array and co-prime array, which can obtain more continuous fourth-order lags than that of nested array [11] and co-prime array [10], respectively.

However, even with the sparse arrays [9–12], the element spacing between the two closest elements still can't exceed the half-wavelength of the incident signal. In practical application, there remains a strong mutual coupling effect between the two elements which can affect the performance of DOA estimation algorithms. Admittedly, there are also some large-spacing sparse arrays [13, 14] with element spacing not limited by half-wavelength. However, since the co-array concept is diffcult to be used, no virtual element can be used to increase the DOFs.

In this work, to reduce the mutual coupling between sensors, a novel sparse array with increasing spacing has been proposed. The element spacing between the two closest elements can be adjusted to an arbitrary integer multiple of the half-wavelength of the incident signal. In a small number of elements, the fourth-order different co-array (FODCA) of the proposed array can even include more continuous fourth-order lags than that of sparse arrays [9–12].

2 Receive Signal Model

Consider that K far-field, non-Gaussian and narrowband signals are received by an M-element linear array. The kth signal is represented as $s_k(t)$, $k = 1, 2, \cdots, K$, and the observed signal from the mth sensor can be expressed as

$$x_m(t) = \sum_{k=1}^{K} e^{-i\frac{2\pi d_m \sin\theta_k}{\lambda}} s_k(t) + n_m(t), m = 1, 2, \cdots, M \quad (1)$$

where d_m is the distance between the reference element and the mth sensor, θ_k is the DOA of the kth signal, λ is the wavelength of the incident signal and $n_m(t)$ is the Gaussian noise received by the mth sensor.

Using the received signals $x_u(t), x_v(t), x_h(t), x_l(t)$, we can define the fourth-order cumulant as [12]

$$c_{u,v,h,l} = cum\{x_u(t), x_v^*(t), x_h(t), x_l^*(t)\} \quad (2)$$

where $()^*$ is conjugate operation.

Using the M physical sensors, we can get a FODCA with the position set as [12]

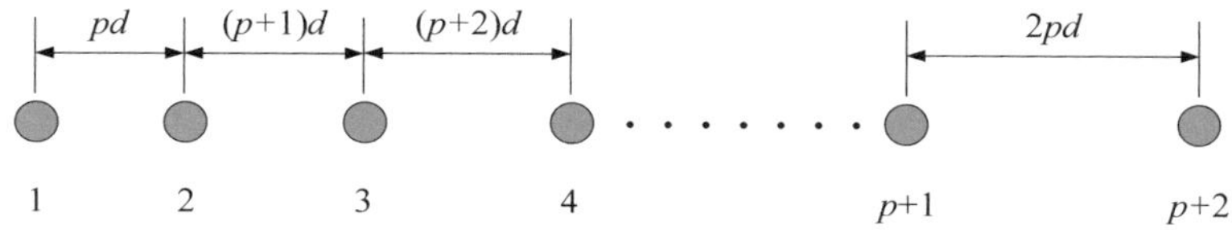

Fig. 1 Structure of the proposed array

$$DCA_f = \{d_u - d_v + d_h - d_l | u, v, h, l = 1, 2, \cdots, M\}. \tag{3}$$

The elements in DCA_f correspond to the fourth-order cumulants one by one. Using the fourth-order cumulants that correspond to the continuous fourth-order lags in DCA_f, we can get a non-redundancy fourth-order cumulant vector. The fourth-order cumulant vector can be segmented and reconstituted into an extended fourth-order cumulant matrix as [7]. Then, the DOAs of all signals can be estimated by some subspace algorithms.

3 The Proposed Sparse Array

3.1 Arrangement of the Sensors

Suppose that the number of the sensors of the proposed array is $M = n + 1$. The initial spacing for the array is pd and the element spacing increases with d as the tolerance, where $p = n - 1$ and $d = \frac{\lambda}{2}$. The structure of the array is shown in Fig. 1, the position set of the proposed array is given as

$$D_p = \{0, pd, (2p + 1)d, \ldots, (\frac{2p + (n - 1)}{2} n)d\} \tag{4}$$

and the position of the mth physical sensor is

$$\frac{2p + (m - 2)}{2}(m - 1)d, m = 1, 2, \ldots, M. \tag{5}$$

Obviously, the element spacing of the proposed array increase as n, so the mutual coupling effect can decrease with the increase of n.

3.2 FODCA for the Proposed Array

We first construct the non-negative second-order different co-array (SODCA) as

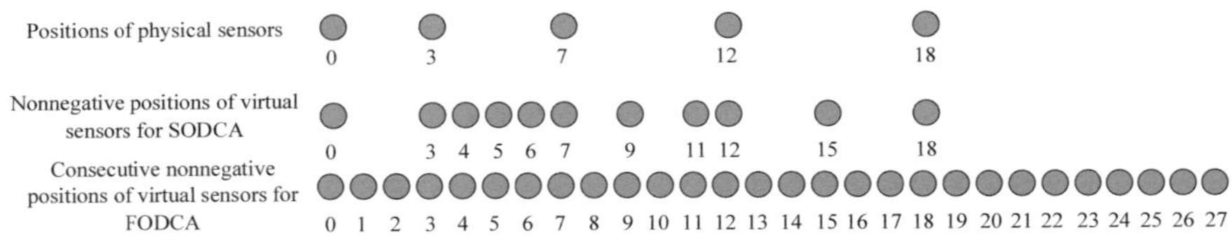

Fig. 2 Structure of a 5-element sparse array

$$C_t^+ = \{d_u - d_v | d_u \geq d_v\,,\, |d_u, d_v \in D_p\}. \quad (6)$$

It's easy to know that C_t^+ can be expressed as

$$C_t^+ = \{0\} \cup C_{t,1}^+ \cup \cdots \cup C_{t,n}^+, \quad (7)$$

where $C_{t,s}^+$ contains $n+1-s$ elements and has the form as

$$C_{t,s}^+ = \left\{ \begin{array}{l} \frac{2p+(s-1)}{2}sd,\ \left[\frac{2p+(s-1)}{2}s+s\right]d,\ldots, \\ \left[\frac{2p+(s-1)}{2}s+(n-s)s\right]d \end{array} \right\},\quad s = 1, 2, \ldots, n. \quad (8)$$

The FODCA for the proposed array can be obtained by

$$C_f = \{a \pm b | a\,,\, b \in C_{t,s}^+\}. \quad (9)$$

Often, we mainly consider the continuous fourth-order lags in C_f. Figure 2 shows the non-negative SODCA and the FODCA of a 5-element sparse array, where the unit length d is deleted for brevity. Table 1 shows the comparison of the number of the continuous fourth-order lags for five different sparse arrays. From the comparison result, we can find that the proposed sparse array shows good performance in generating continuous fourth-order lags for a small number of sensors.

4 Simulation Experiment

In this section, we compare the performance of the proposed sparse array with the other three arrays in DOA estimation. For all sparse arrays, the DOAs are estimated by dealing with the fourth-order cumulant matrix in the method [7]. We suppose that the number of the sensors is five for the four kinds of sparse array.

In the first experiment, we suppose that the mutual coupling effect has been ignored. Consider six signals with the DOAs as $[-10°, 0°, 10°, 20°, 30°, 40°]$. Figure 3 shows the comparison of spatial spectrum for the four arrays with SNR =

Table 1 Comparison of the number of the continuous fourth-order lags for five different sparse arrays

	5 sensors	6 sensors	7 sensors	8 sensors	9 sensors	10 sensors
Proposed array	55 ($p = 3$)	91 ($p = 4$)	143 ($p = 5$)	187($p = 6$)	261($p = 7$)	285($p = 8$)
Four-level nested array [9]	47 $N_1 = 2$ $N_2 = 2$ $N_3 = 2$ $N_4 = 2$	69 $N_1 = 3$ $N_2 = 2$ $N_3 = 2$ $N_4 = 2$	107 $N_1 = 3$ $N_2 = 3$ $N_3 = 2$ $N_4 = 2$	161 $N_1 = 3$ $N_2 = 3$ $N_3 = 3$ $N_4 = 2$	215 $N_1 = 3$ $N_2 = 3$ $N_3 = 3$ $N_4 = 3$	287 $N_1 = 4\, N_3 = 3$ $N_2 = 3\, N_4 = 3$
Coprime array [10]	Inexistence	Inexistence	Inexistence	139 $N_1 = 2$ $N_2 = 3$ $N_3 = 2$	189 $N_1 = 2$ $N_2 = 3$ $N_3 = 3$	241 $N_1 = 2$ $N_2 = 3$ $N_3 = 4$
Expanded nested array [11]	53 $N_1 = 1$ $N_2 = 2$ $N_3 = 2$	85 $N_1 = 2$ $N_2 = 2$ $N_3 = 2$	133 $N_1 = 2$ $N_2 = 3$ $N_3 = 2$	183 $N_1 = 2$ $N_2 = 3$ $N_3 = 3$	249 $N_1 = 3$ $N_2 = 3$ $N_3 = 3$	337 $N_1 = 3$ $N_2 = 4$ $N_3 = 3$
Expanded nested array [12]	55 $M_1 = 1$ $M_2 = 2$ $N_1 = 1$ $N_2 = 2$	87 $M_1 = 2$ $M_2 = 2$ $N_1 = 1$ $N_2 = 2$	135 $M_1 = 2$ $M_2 = 3$ $N_1 = 1$ $N_2 = 2$	203 $M_1 = 2$ $M_2 = 3$ $N_1 = 2$ $N_2 = 2$	305 $M_1 = 2$ $M_2 = 3$ $N_1 = 2$ $N_2 = 3$	413 $M_1 = 3$ $M_2 = 3$ $N_1 = 2$ $N_2 = 3$

10 dB and snapshots being 500. The result shows that all signals can be distinguished clearly with the four different arrays with mutual coupling being ignored.

In the second experiment, we suppose that the mutual coupling only exists between the sensors that are close together. The mutual coupling coefficient is $0.4 + 0.3i$ for the two sensors with spacing being d. The mutual coupling coefficient is 0 for the two sensors with spacing being larger than d. We first consider three signals with the DOAs as $[10°, 20°, 30°]$. Figure 4 shows the comparison of spatial spectrum for the four arrays with SNR = 10 dB and snapshots being 500. Then we consider four signals with the DOA as $[10°, 20°, 30°, 40°]$. Figure 5 shows the comparison of spatial spectrum for the four arrays with SNR = 10 dB and snapshots being 500. By combining the two figures, we can see that the mutual coupling effect will affect the performance of the algorithm when the number of signals is larger.

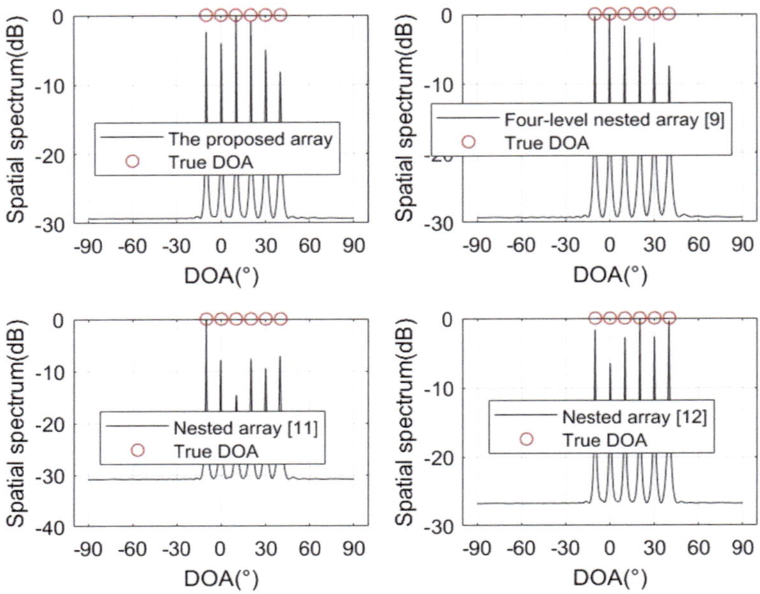

Fig. 3 Comparison of spatial spectrum by the four arrays with mutual coupling being ignored

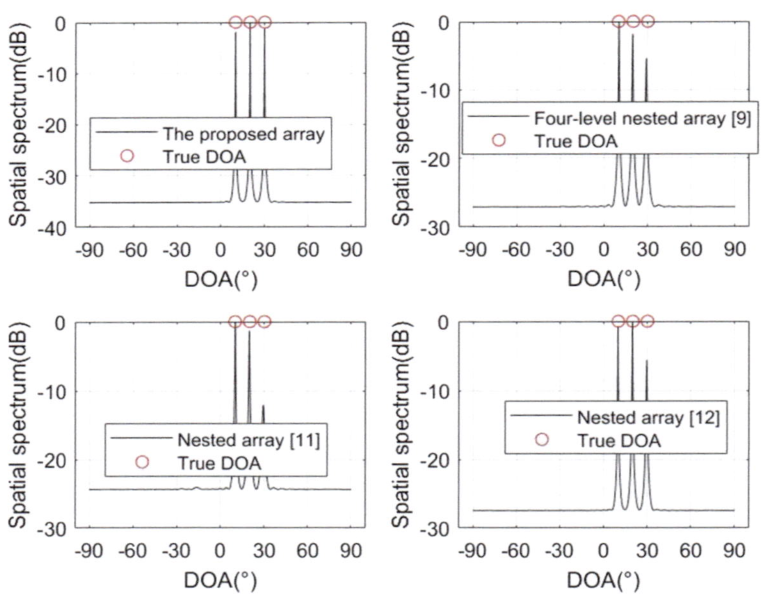

Fig. 4 Comparison of spatial spectrum by the four arrays with mutual coupling being considered for three signals

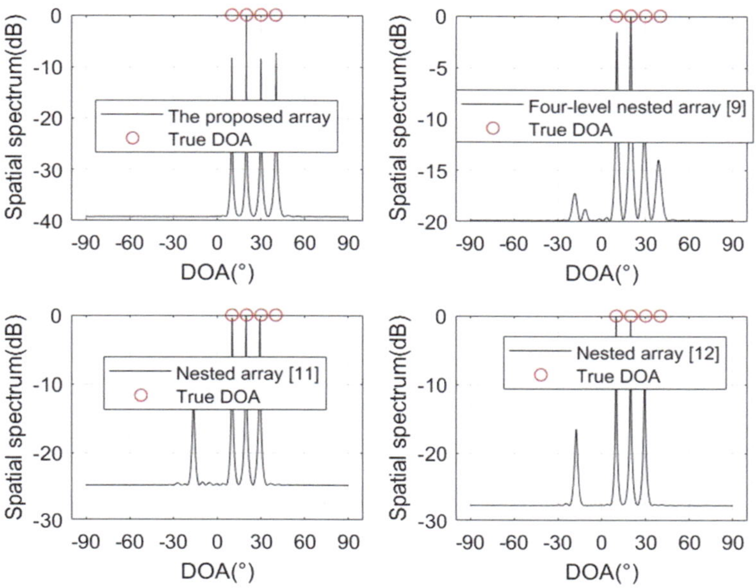

Fig. 5 Comparison of spatial spectrum by the four arrays with mutual coupling being considered for four signals

5 Conclusion

A sparse array with increasing spacing is proposed for DOA estimation. The spacing of any two adjacent elements can increase as the number of sensors. Compared with many other kinds of sparse arrays, the outstanding advantage of the proposed array is the strong robustness to the mutual coupling effect. In addition, when the number of sensors is small, the proposed array can even generate more continuous fourth-order lags than some existing sparse arrays.

Acknowledgements This work was supported by the National Natural Science Foundation of China (62261047), the Science and Technology Foundation of Guizhou Province of China (ZK[2022]557), the Tongren Science and Technology Planning Project ([2022]59) and the Science and Technology Innovation Team of Guizhou Education Department (Grant No. [2023]094)

References

1. Vaidyanathan, P.P., Pal, P.: Sparse sensing with co-prime samplers and arrays. IEEE Trans. Signal Process. **59**(2), 573–586 (2010)
2. Pal, P., Vaidyanathan, P.P.: Nested arrays: a novel approach to array processing with enhanced degrees of freedom. IEEE Trans. Signal Process. **58**(8), 4167–4181 (2010)

3. Ren, S.W., Wang, W.J., Chen, Z.H.: DOA estimation exploiting unified coprime array with multi-period subarrays. In: 2016 CIE International Conference on Radar (RADAR), pp. 1–4. IEEE, Guangzhou, China (2016).
4. Chen, M., Gan, L., Wang, W.: Co-prime arrays with reduced sensors (CARS) for direction-of-arrival estimation. In: 2017 Sensor Signal Processing for Defence Conference (SSPD), pp. 1–4. IEEE, London, UK (2017)
5. Yang, M., Sun, L., Yuan, X., Chen, B.: Improved nested array with hole-free DCA and more degrees of freedom. Electron. Lett. **52**(25), 2068–2070 (2016)
6. Iizuka, Y., Ichige, K.: Extension of nested array for large aperture and high degree of freedom. IEICE Commun. Express **6**(6), 381–386 (2017)
7. Liu, S., Liu, Q., Zhao, J., Yuan, Z.: Triple two-level nested array with improved degrees of freedom. Progress Electromagnetics Res. **84**, 135–151 (2019)
8. Liu, S., Cao, H., Wu, D., Chen, X.: Generalized array architecture with multiple sub-arrays and hole-repair algorithm for DOA estimation. Comput. Mater. Continua **64**(1), 589–605 (2020)
9. Pal, P., Vaidyanathan, P.P.: Multiple level nested array: an efficient geometry for 2qth order cumulant based array processing. IEEE Trans. Signal Process. **60**(3), 1253–1269 (2011)
10. Shen, Q., Liu, W., Cui, W., Wu, S.: Extension of co-prime arrays based on the fourth-order difference co-array concept. IEEE Signal Process. Lett. **23**(5), 615–619 (2016)
11. Shen, Q., Liu, W., Cui, W., Wu, S.: Extension of nested arrays with the fourth-order difference co-array enhancement. In: 2016 IEEE International Conference on Acoustics, Speech and Signal Processing (ICASSP), pp. 2991–2995. IEEE, Shanghai, China (2016)
12. Cai, J., Liu, W., Zong, R., Shen, Q.: An expanding and shift scheme for construct-ing fourth-order difference co-arrays. IEEE Signal Process. Lett. **24**(4), 480–484 (2017)
13. Liu, S., Zhao, J., Wu, D., Zeng, C., Mao, Z., Huang, Y.: 2D DOA estimation by a large-space T-Shaped array. Digital Signal Processing **130**, 103699 (2022)
14. Liu, S., Zhao, J., Wu, D., Huang, Y., Xia, L.: An unambiguous 2D DOA estimation algorithm by a large-space L-shaped array. Circ. Syst. Signal Process. **42**, 6614–6635 (2023)

An Img2Vec Based Floor Recognition and Implementation Method

Zhiqing Zhang, Jing Li, Liang Yan, Shulin Wu, and Xu Zhang

Abstract Understanding floor image data is important for planning, 3D construction. In this paper, we propose a novel approach for floor content recognition and implementation with Img2Vec, which is converts images into vectors. The proposed method can obtain better performance with 3 key steps: character erasion, corner detection, corner re-order. The implemented system can process input floor image into vector file and corresponding coordinate to generate a certain polygon.

Keywords Image to vector · Harris corner · Floor recognition

1 Introduction

Recent years have witnessed remarkable strides in image representation learning, yielding notable successes across diverse downstream tasks like image reconstruction and classification [1, 2]. However, the majority of these advancements have predominantly focused on analyzing structured bitmap formats, leaving significant room for improvement in representing detailed attributes for vector objects. Unlike bitmap images, scalable vector graphics (SVG) formats, with their inherent scalability and ability to represent intricate curves and features, offer immense potential for real-world applications [2, 3].

In industrial design contexts, vector graphics, particularly in 2D floorplans, serve as foundational representations for interior design and property development. Despite

Z. Zhang · J. Li · L. Yan
Chongqing Planning and Natural Resources Information Center, Chongqing, China

Key Laboratory of Monitoring, Evaluation and Early Warning of Territorial Spatial Planning Implementation, Ministry of Natural Resources, Chongqing, China

S. Wu (✉) · X. Zhang
Department of Computer Science and Technology, Chongqing University of Posts and Telecommunications, Chongqing, China
e-mail: 1256829298@qq.com

J. S. Park et al. (eds.), *Advanced Multimedia and Ubiquitous Engineering*, Lecture Notes in Electrical Engineering 1475, https://doi.org/10.1007/978-981-95-1565-3_34

their advantages over raster images, the irregular structure of vector graphics complicates tasks such as semantic segmentation, essential for applications like room layout recognition. Automatic recognition of room layouts from wall structures in floorplans presents a particularly challenging task, necessitating precise semantics and boundary delineation. In this paper, we propose a novel methodologies tailored to address the unique characteristics of SVGs [3, 4], bridging the gap between traditional image representation learning and the complexities of vector graphics. By leveraging advanced machine learning techniques, we achieve an outstanding performance in understanding image to vector.

2 Method and Implementation

2.1 Key Point Extraction Using Corner Detection

To identify key points in the floor images, we employ the Harris corner detector, renowned for its robustness and accuracy. The proposed method involves several key steps:

(1) Preprocessing: The input image undergoes preprocessing steps such as grayscale conversion and Gaussian smoothing to reduce noise. Utilizing Sobel or Scharr operators, the image gradient is computed to capture intensity changes. Then, the structure tensor is computed to characterize the local image structure based on the gradients.

(2) Corner Response Calculation: The Harris corner response function evaluates the likelihood of each pixel being a corner based on the eigenvalues of the structure tensor. It is defined as:

$$R = \lambda_1 \lambda_2 - k(\lambda_1 + \lambda_2)^2$$

where $\lambda_1 \lambda_2$ are the eigenvalues of the structure tensor, and k is a constant typically set to a small value.

To enhance robustness against noise, a thresholding mechanism is employed to filter out weak corners. Finally, corner coordinates are localized and extracted from the image, representing the key points of interest. The Psycode is illustrated in Algorithm 1.

Algorithm 1: Psycode for Img2Vec data

Input: floor image; Output: detected corner coordinates

1. Compute Gradient:
   Compute horizontal gradient I_x and vertical gradient I_y
2. Build Covariance Matrix:
   Compute squared gradients and cross-product of gradients:
3. Weighted Summation:
   Apply weighted window w (e.g., Gaussian or average weighted) to compute weighted sums of squared gradients and cross-product:
   $a = w * {I_x}^2$, $b = w * I_{xy}$, $c = w * {I_y}^2$
4. Compute Eigenvalues of Covariance Matrix:
   For each pixel in the image:
   Compute the covariance matrix elements:
   M = | a b |
   | b c |
   Compute the eigenvalues λ_1 and λ_2 of M
   Compute the Harris corner response R using the formula:
   $R = \lambda_1 \lambda_2 - k(\lambda_1 + \lambda_2)^2$
5. Corner Detection:
   Keep pixels with corner response $R > \theta$ as corner points

Output: Detected corner points based on the specified threshold

2.2 *Polygon Generation*

Simple use obtained corner points, we can not define a certain polygon. Therefore, we follow the floor image as a guide to calculate the polygon. We adopt a snake-based method inspired by active contour models, integrated with affine transformation:

(1) Initialization: Before contour evolution, an affine transformation is applied to the image to align it with the snake contour, facilitating better convergence towards the polygon edges.
(2) Contour Evolution: The snake contour evolves iteratively to minimize the overall loss function, adapting its shape to conform to the polygon edges. The loss function comprises internal and external energy terms, incorporating gradient information from the image to guide the contour evolution. The evolution process continues until convergence, indicating stable alignment of the snake contour with the polygon edges.
(3) Vertex Extraction: Upon convergence, vertices of the sorted polygon are extracted from the snake contour, providing the ordered sequence of points defining the polygon boundary.

The extracted key points from corner detection are matched with the sorted vertices obtained from the snake-based polygon sorting. This integration process aligns key points with corresponding vertices, facilitating further geometric processing of the polygon.

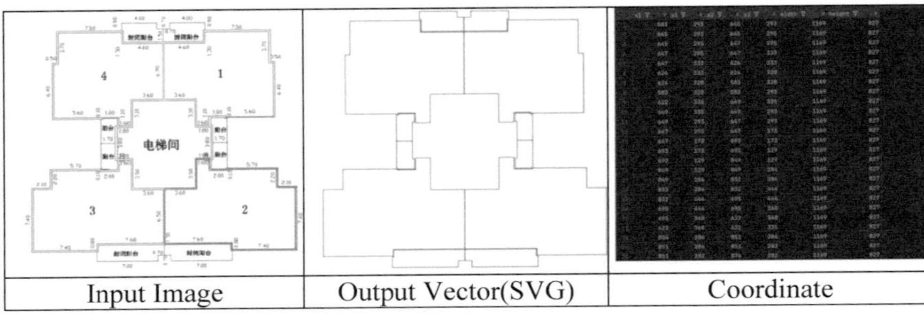

Fig. 1 Image to vector recognition demonstration

2.3 Implementation Result

The implemented system can dynamic import certain floor plan, erase the characters in the image and transform into a vector file. As shown in Fig. 1, the output include a SVG file and certain coodinates of each corner representing a polygon.

3 Conclusion

In this paper, we developed an img2vec based floor recognition method. The implemented system can process floor image and vectorize into SVG and polygon coordinates, which can effectively save large volume of human resource for labeling and redrawing floor vector data.

References

1. Cao, D., et al., Svgformer: representation learning for continuous vector graphics using transformers. In: Proceedings of the IEEE/CVF Conference on Computer Vision and Pattern Recognition (2023)
2. Liu, Y.T., et al., Dualvector: unsupervised vector font synthesis with dual-part representation. In: Proceedings of the IEEE/CVF Conference on Computer Vision and Pattern Recognition, pp. 14193–14202 (2023)
3. Yang, B., et al., Vectorfloorseg: two-stream graph attention network for vectorized roughcast floorplan segmentation. In: Proceedings of the IEEE/CVF Conference on Computer Vision and Pattern Recognition, pp. 1358–1367 (2023)
4. Song, J.Q., Su, F., Chen, J.B., Cai, S.J.: An orientation-independent linear vectorization method based on seed segments. J. Softw. **11**(9), 1196–1201 (2000). (In Chinese)

A Multi Label Based Classification Method for Commodity Risk Assessment

Hongyu Tu, Houbing Zhang, Changjiu Ke, Yuran Dai, and Xu Zhang

Abstract It is a critical problem to identify the risk in imported or exported commodities. In this paper, we presents a multi-label-based classification approach for risk assessment. Firstly, we introduce the labeling procedure for defects description in the collected data. Furthermore, we introduce a novel risk scoring mechanism and GRU module to quantify the severity of identified risk labels, facilitating prioritization of risk mitigation strategies. Finally, we classify with national standard and implement a system in Chongqing.

Keywords Multi-lable classification · GRU · Risk assessment

1 Introduction

In the dynamic landscape of import and export commodity markets, assessing and managing risks associated with commodity trading stands as a formidable challenge [1, 2]. In recent years, advancements in artificial intelligence (AI) and machine learning (ML) have offered promising avenues for enhancing risk assessment strategies [3]. In this paper, we introduces a novel approach for commodity risk assessment leveraging a multi-label-based classification method. Unlike traditional binary classification models that categorize assets into discrete risk categories, our method recognizes the multi-dimensional nature of commodity risk by assigning multiple risk labels to each asset. The proposed method captures the nuanced interplay of various risk factors and provides a more comprehensive understanding of the risk landscape.

In this paper, we propose an data enhancement and multi-label prediction method based on hierarchical analysis and matrix analysis. Firstly, we carry out entity

H. Tu · H. Zhang · C. Ke
Chongqing Customs Technology Center, Chongqing, China

Y. Dai (✉) · X. Zhang
Department of Computer Science and Technology, Chongqing University of Posts and Telecommunications, Chongqing, China
e-mail: 2511052698@qq.com

J. S. Park et al. (eds.), *Advanced Multimedia and Ubiquitous Engineering*, Lecture Notes in Electrical Engineering 1475, https://doi.org/10.1007/978-981-95-1565-3_35

identification, emotion analysis and risk grading of defect description information commodity text data. Then, we implement a multi-label super network to train and learn risk featuers. Finally, the risk assessment is evaluated based on risk matrix mentioned in GB/T 22760-2020 [4].

2 Methodology

2.1 Labeling

Through the existing customs import and export of children's clothing merchandise data defects and consequences of data cleaning, that is, pre-processing operations for lexicaling, and its removal of disabled words and illegal characters. For example, for the sentence "The recalled jackets have drawstrings in the hood and drawstrings in the waist area that can wrap around or hang from playground slides, handrails, school bus doors, or other moving objects, posing a strangulation and/or entrapment hazard to children. Playground jacket door strangulation wrap harness object slide child danger hang handrail would pinch recall hat area move cause school bus". Then we can run the word frequency statistics and then combine the keywords labelled by humans to extract some keywords related to the classification labels, such as "harness", "entanglement", etc., and then we can target the unqualified items such as "harness"; "safety"; "safety"; "burns"; "safety", etc. Safety", "burns", "flammability" and other keywords, we can target their failure as "combustion performance".

2.2 Multi-Label Classification

For commodity risk classification, due to the occurrence of multiple labels, for example, for the sentence "phthalate exposure to plastics may damage the liver and kidneys, exposure to formaldehyde can lead to vision and skin diseases, digestive and respiratory disorders," it will have two labels for phthalate and formaldehyde content. It would have two labels for phthalates and formaldehyde content. So essentially this is a multi-label classification problem.

Firstly, we analyze the data as a whole, using the maximum data length of 200, and then normalize the original data to a specified length, and then use these processed data to extract textual features using ALBERT, which improves the learning effect in the case of reducing the number of parameters.

$$z_t = \sigma\left(W_z \cdot \left[h_{t-1}, x_t\right]\right) \tag{1}$$

$$r_t = \sigma\left(W_r \cdot \left[h_{t-1}, x_t\right]\right) \tag{2}$$

$$\tilde{h}_t = \tanh\left(W \cdot \left[r_t * h_{t-1}, x_t\right]\right) \tag{3}$$

$$h_t = (1 - z_t) * h_{t-1} + z_t * \tilde{h}_t \tag{4}$$

The GRU is involved with the detailed calculation exhibited in Eq. (1)–(4). After obtained the text features via ALBERT, it will be processed with a GRU layer, and the corresponding formula is shown in Eq. 1.The input–output structure of the GRU [5] is the same as that of an ordinary RNN, i.e., the input xt at the current moment and the hidden state h_{t-1} passed down from the previous node, which contains the relevant information of the previous node, and then combining the xt and h_{t-1}, the GRU will get the output yt of the current hidden node and the hidden state h_t passed down to the next node, which can solve the problem of memorization for long sentences.

3 Implementation

We implement the propose method in Chongqing custom commodity risk assessment center. The implemented system can dynamic capture recent commodity description, risk descriptions in the world. Then, the data can be processed with the proposed method to obtain a certain risk level and output a formal report for decision support. The collected commodity information are shown in Fig. 1.

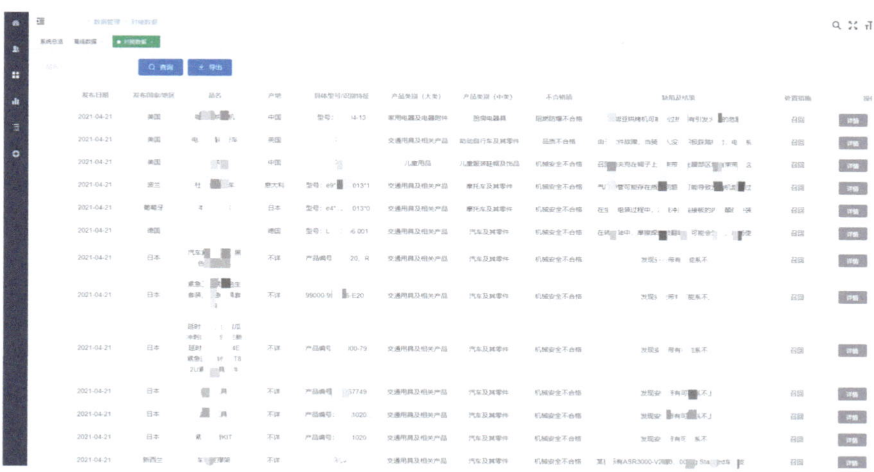

Fig. 1 Risk assessment system in Chongqing

4 Conclusion

In this paper, we developed a risk assessment method based on multi-label classification, and provide technical support for the commodity risk assessment of customs imported children's clothing in Chongqing. The implemented system demonstrate a good performance in data collection and risk assessment.

References

1. Mou, J., Cai, J., Yu, M., Xu, J.: A multi-label classification algorithm for class attributes based on label correlation. Comput. Appl. Res. **37**(9), 4 (2020). https://doi.org/10.19734/j.issn.1001-3695.2019.04.0118. (In Chinese)
2. Xu, B., Lu, W., Li, H., et al., Risk identification and level determination in risk assessment of import and export commodities. Insp. Quar. Sci. (02), 58–61 (2008). CNKI:SUN:XDSJ.0.2008-02-022. (In Chinese)
3. Wever, M., et al.: AutoML for multi-label classification: overview and empirical evaluation. IEEE Trans. Pattern Anal. Mach. Intell. **43**(9), 3037–3054 (2021)
4. 《Consumer product safety-General principles for risk assessment》, GB/T 22760-2020
5. Ruan, X., et al., Real-time risk prediction of colorectal surgery-related post-surgical complications using GRU-D model. J. Biomed. Inf. **135**, 104202 (2022)

Comparative Learning Based Multi-round Dialogue Intent Classification Method

Feng Wei, Chenzi Wang, Yuan Huang, and Xu Zhang

Abstract The traditional text classification method faces great challenges when processing with the multi-round dialogue including several intents. In this paper, we propose an intent classification model based on comparative learning and an attention mechanism. The text is divided into long and short categories and encoded into the Transformer. Then, the word embedding matrix is perturbed to generate adversarial samples, and the positive sample pairs are then compared for loss. Finally, the positive sample pair is input into the multi-round inference module, and the inference features are obtained by learning the semantic clues in the whole scene through multi-round dialogue. Experiments on two datasets exhibit that the proposed method achieved good performance.

Keywords Dialogue classification · Comparative learning · Text classification

1 Introduction

The advent of natural language processing technology has precipitated the widespread utilization of text classification techniques across a plethora of disciplines. In recent years, online customer service has utilized a significant number of text classification methods with a view to enhancing the efficiency and quality of service. Conversely, traditional online customer service relies on text communication and has achieved favorable outcomes in dialogue classification. However, voice customer service necessitates manual classification, which is both inefficient and costly. Multi-round dialogue intent classification in voice customer service scenarios poses significant challenges, including word repetitions, unclear intentions, and ultra-long text content. The existing classification models are inadequate in addressing the

F. Wei · C. Wang · Y. Huang
China Telecom Corporation Limited Chongqing Branch, Chongqing, China

X. Zhang (✉)
School of Artificial Intelligence, Chongqing University of Posts and Telecommunications, Chongqing, China
e-mail: zhangx@cqupt.edu.cn

J. S. Park et al. (eds.), *Advanced Multimedia and Ubiquitous Engineering*, Lecture Notes in Electrical Engineering 1475, https://doi.org/10.1007/978-981-95-1565-3_36

aforementioned issues. Recent advances in deep learning and neural network theory have demonstrated the potential ability for enhanced text classification including Bert [1] and DPCNN [2]. Text classification models based on Convolutional Neural Networks have seen significant development, with TextCNN utilizing convolution to capture n-grams in sentences [3] and capturing important features in the text through maximum pooling while addressing the issue of inconsistent sentence length. Pang et al. [4] have proposed a heterogeneous graph-based method to encode the whole dialogue by explicitly modelling multiple factors that are crucial for cross-lingual, speaker-specific and contextual information extraction. In this paper, we proposes a comparative learning method that aims to improve classification performance in complex multi-round dialogue scenarios.

2 Methodology

The proposed method mainly consists of a comparative learning module, a multi-round scenario inference module and a contrast loss function, as shown in Fig. 1.

The input utterances are categorized into long/short sequences through adaptive threshold. Each sequence undergoes positional encoding and is processed through a Transformer encoder with multi-head attention, generating contextualized representations $H \in H \in R^{d \times n}$ where d denotes embedding dimension and n sequence length. We implement manifold mixup regularization by applying linear perturbations to the word embedding matrix, which generates adversarial sample pairs through stochastic feature space interpolation while maintaining semantic consistency.

Then, we add perturbations to the word embedding matrix directly on transformer-based encoders to generate adversarial samples. Then, the word embedding matrix V obtained by calculating its gradient through the classifier loss replaces the word embedding to calculate the perturbations, and the loss function used by the classifier is the cross-entropy loss function:

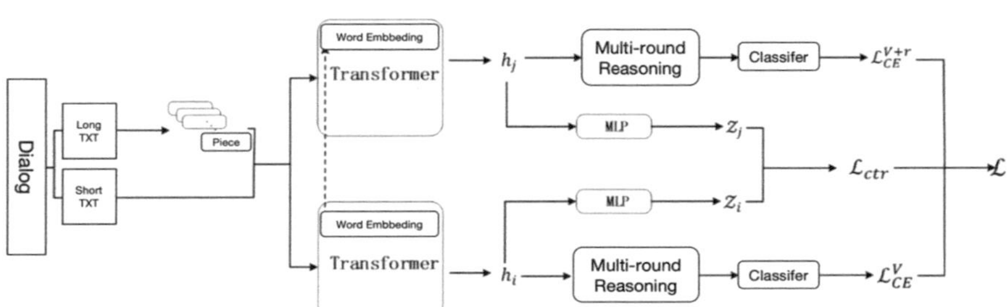

Fig. 1 The proposed architecture

$$\mathcal{L}_{CE} = -\frac{1}{N}\sum_{i=1}^{N}\sum_{c=1}^{C} y_{i,c}\log\left(p\left(y_{i,c}h^{i}_{[CLS]}\right)\right) \quad (1)$$

Based on the idea of comparative learning, the sample gets the hidden vector of the last layer of the transformer $h^{i}_{[CLS]}$ and the one obtained from the antagonistic sample as a pair of positive samples $h^{j}_{[CLS]}$, and then a nonlinear projection layer is added:

$$z_i = W_2 ReLU\left(W_1 h^{i}_{[CLS]}\right), z_j = W_2 ReLU\left(W_1 h^{j}_{[CLS]}\right) \quad (2)$$

There are 2(N-1) negative sample pairs for each positive sample pair in a batch of N sample pairs. The contrast loss is defined as follows:

$$\mathcal{L}_{ctr} = -\log\frac{exp\left(cossimilar\left(z_i, z_j/\tau\right)\right)}{\sum_{k=1}^{2N} 1_{[k\neq i]} exp\left(cossimilar\left(z_i, z_j/\tau\right)\right)} \quad (3)$$

3 Experimental Evaluation

To evaluate the performance of the proposed method, we did experiments on the public dataset DailyDialog and a private dataset China Telecom. The experimental evaluations are exhibited in Tables 1 and 2.

As shown in Table 1, the proposed method outperform most of the methods and achieves comparative results as MFDG. Then, we conduct experiments and on a

Table 1 Results on DailyDialog

Model	Acc	Top-3	Top-5	Macro-F1
TextRNN	49.10	74.90	74.00	38.62
Bert-base [1]	62.90	83.40	88.60	57.48
DialogueGCN [5]	61.30	83.90	90.80	52.48
MFDG [4]	**66.50**	89.30	94.40	60.64
Ours	65.98	**89.42**	**94.88**	**61.03**

Table 2 Results on China Telecom (10,000)

Class	P (Top1)	R	F1	Top2	Top3	Top5	Acc	Support
Apply	0.820	0.811	0.813	0.932	0.968	0.985	0.890	1132
Query	0.653	0.649	0.649	0.749	0.788	0.819	0.680	2029
Fault	0.689	0.666	0.672	0.795	0.841	0.855	0.756	1691
Complain	0.620	0.608	0.610	0.688	0.735	0.748	0.750	3320

private dataset provided by China Telecom Chongqing, introduces Top k accuracy assessment method and three indicators, Top 2, Top 3 and Top 5 as shown in Table 2. For the private dataset, the original multi-round dialogue are text recognized from the original dialogue audio. There are a lot of repetition, wrong descriptions along with multi-intents. For example, the user may describe their query request in the beginning and begin to complain after several seconds. The original dialogue text are encoded directly into the proposed method and generate the classified intent with four categories including Apply, Query, Fault and Complain.

4 Conclusion

In this paper, we proposed a comparative learning method for intent classification with multi-round conversations. The experiments on two datasets exhibit a good performance in real environments.

References

1. Aboutaleb, A.: BERT BiLSTM-attention similarity model. In: 2021 IEEE International Conference on Artificial Intelligence and Computer Applications (ICAICA), pp. 88–93. IEEE (2021)
2. Johnson, R., Tong, Z.: Deep pyramid convolutional neural networks for text categorization. In: Proceedings of the 55th Annual Meeting of the Association for Computational Linguistics, vol. 1, pp. 97–102 (2017)
3. Wang, H., He, J., Zhang, X., et al.: A short text classification method based on N-gram and CNN. Chin. J. Electron. **29**(2), 248–254 (2020)
4. Pang, J., Xu, H., Song, S., et al.: MFDG: a multi-factor dialogue graph model for dialogue intent classification. ECML PKDD **13714**, 691–706 (2022)
5. Yan, Y., Zhang, B.W., Min, P., et al.: DialGNN: heterogeneous graph neural networks for dialogue classification. Neural. Process. Lett. **56**(2), 143 (2024)

MIX
Papier aus verantwortungsvollen Quellen
Paper from responsible sources
FSC® C105338

If you have any concerns about our products,
you can contact us on
ProductSafety@springernature.com

In case Publisher is established outside the EU,
the EU authorized representative is:
Springer Nature Customer Service Center GmbH
Europaplatz 3, 69115 Heidelberg, Germany

Printed by Libri Plureos GmbH
in Hamburg, Germany